Multinational Agribusinesses

Ruth Rama, PhD
Editor

Routledge
Taylor & Francis Group
New York London

First published 2005 by The Haworth Press, Inc.,

Published 2021 by Routledge
605 Third Avenue, New York, NY 10017
2 Park Square, Milton Park, Abingdon, Oxon OX14 4RN

Routledge is an imprint of the Taylor & Francis Group, an informa business

Cover design by Jennifer M. Gaska.

Library of Congress Cataloging-in-Publication Data

Multinational agribusinesses / Ruth Rama, editor.
      p. cm.
   Includes bibliographical references and index.
   ISBN 1-56022-936-5 (hard : alk. paper) — ISBN 1-56022-937-3 (soft : alk. paper)
      1. International business enterprises.   2. Globalization—Economic aspects.   3. International economic relations.   4. Agricultural industries—Management.   5. Food industry and trade. 6. Investments, Foreign.   I. Rama, Ruth.

HD2755.5M817 2004
338.8'873—dc22

                                                                          2004005924

ISBN 13: 978-1-56022-936-0 (hbk)

CONTENTS

About the Editor   ix

Contributors   xi

Acknowledgments   xiii

Introduction   xv
     Ruth Rama

Chapter 1. The Rise of Global Enterprises in the World's Food Chain   1
     Selma Tozanli

Introduction	1
A Dichotomous Globalization Process	3
Sectoral Globalization: Refocusing on Core Business Activities	15
The Changing Structure of the World's Food Oligopoly	22
Concluding Remarks	27
Appendix	28

Chapter 2. The Performance of Multinational Agribusinesses: Effects of Product and Geographical Diversification   73
     George Anastassopoulos
     Ruth Rama

Introduction	73
Theoretical Background	76
Diversification, Size, and Rate of Growth	80
Econometric Specifications	94
Concluding Remarks	104
Appendix	107

Chapter 3. Innovation in Food and Beverage Multinationals   **115**

> *Oscar Alfranca*
> *Ruth Rama*
> *Nicholas von Tunzelmann*

Introduction   115
Patented Innovation in Food and Beverage Multinationals   117
The Globalization of Innovative Activities   120
Innovators and Their Strategies   131
Concluding Remarks   138
Appendix   142

Chapter 4. Foreign Direct Investment in U.S. Food and Kindred Products   **149**

> *Daniel Pick*
> *Thomas Worth*

Introduction   149
Recent Trends in FDI   150
What Is Foreign Direct Investment and Why?   156
Determinants of FDI   158
Future Research   162

Chapter 5. Multinational Food Corporations and Trade: The Impact of Foreign Direct Investment on Trade in the U.S. Food Industry   **165**

> *Andrew P. Barkley*

Introduction   165
Previous Literature   167
An Empirical Model of the Impact of FDI on
    International Trade   172
Data   174
Regression Results   177
Implications and Conclusions   187

**Chapter 6. External versus Internal Markets
of the Multinational Enterprise: Intrafirm Trade
in French Multinational Agribusiness**   **191**
Emmanuelle Chevassus-Lozza
Jacques Gallezot
Danielle Galliano

The Foundations of Intrafirm Trade   193
Intrafirm Trade in French Agrofood Foreign Trade   198
The Determinants of Intrafirm Trade in the French
    Agrofood Industry: Model and Results   205
Conclusions   214

**Chapter 7. The Internationalization Paths of Australian
and New Zealand Food MNEs**   **219**
Bill Pritchard

Introduction   219
Existing Research on Firm Internationalization   224
The History and Political Economy of Antipodean Food
    Sectors   226
Australia/New Zealand As a Base for Global Branded
    Food Companies: The Foster's Group   231
Using Australia/New Zealand As a Regional Base:
    Coca-Cola Amatil   236
Leveraging Competitive Advantage: The Fonterra Story   242
Conclusions   248

**Chapter 8. Regionalization, Globalization,
and Multinational Agribusiness: A Comparative
Perspective from Southeast Asia**   **253**
David Burch
Jasper Goss

Introduction   253
Approaches to the Study of TWMNEs   254
Agrofood Multinational Companies in Asia: Unwieldy
    Giants or Nimble Opportunists?   258
The Charoen Pokphand Group   260
The Salim Group   265

Southeast Asian Agrofood Multinationals: A Comparative
  Analysis   269
Conclusions   276

**Chapter 9. Multinational Firms in the Brazilian Food
Industry**   **283**
  Elizabeth Farina
  Cláudia Viegas

Introduction   283
General Overview of Foreign Direct Investments in Brazil
  and Latin America   286
M&A: Main Strategy of Entry into the Brazilian Market   295
The Brazilian Food Market As the Main Attraction for FDI   298
Impacts on the Food Industry of Brazil   303
Conclusions   307
Appendix   310

Index   **323**

ABOUT THE EDITOR

Ruth Rama, PhD, is Research Professor in the Institute of Economics and Geography at the Spanish Council for Scientific Research in Madrid, where she formerly served as Senior Researcher beginning in 1988. Previously, she was a consultant on agribusiness and foreign direct investment for the Centre on Transnational Corporations of the United Nations (UNCTC), the Food and Agriculture Organization of the United Nations (FAO), and the Organisation for Economic Co-operation and Development (OECD).

Dr. Rama is the author of *Investing in Food* and is widely published in the areas of food and beverage multinationals, and innovation and package design in the food industry. She has served as Guest Editor of the *International Journal of Biotechnology* and is a member of the editorial board of the *International Journal of Entrepreneurship and Innovation Management.*

CONTRIBUTORS

Oscar Alfranca, PhD, is Associate Professor at Universitat Politecnica de Catalunya and Researcher at the Institut Universitari d'Estudis Europeus, Universitat Autonoma de Barcelona, Spain.

George Anastassopoulos, PhD, is Assistant Professor in Enterprise Strategy and Policy, Department of Business Administration, University of Patras, Greece.

Andrew P. Barkley, PhD, is Professor at the Department of Agricultural Economics, Kansas State University.

David Burch, PhD, is Professor at the School of Science, Griffith University, Brisbane, Australia.

Emmanuelle Chevassus-Lozza, PhD, is Director of Research, Institut National de la Recherche Agronomique et Université de Nantes, Unité Mixte de Recherche en Economie-Droit Rural et Agroalimentaire (UMR-EDRA), Nantes, France.

Elizabeth Farina, PhD, is President of the Brazilian Anti-Trust Agency and Professor at the University of São Paulo, Brazil, where she is the Head of the Department of Economics and the Vice-Coordinator of PENSA–Food and Agribusiness Program.

Jacques Gallezot, PhD, is Director of Research, Institut National de la Recherche Agronomique et Institut National Agronomique Paris–Grignon, Unité Mixte de Recherche en Economie Publique, Paris, France.

Danielle Galliano, PhD, is Director of Research, Institut National de la Recherche Agronomique, Unité d'Economie et de Sociologie Rurales, Toulouse, France.

Jasper Goss, PhD, is Researcher Officer at the International Union of Foodworkers (IUF), Asia and Pacific Regional Secretariat, Sydney, Australia.

Daniel Pick, PhD, is Chief of the Specialty Crops Branch, the U.S. Department of Agriculture, Economic Research Service.

Bill Pritchard, PhD, is Senior Lecturer in Economic Geography, Division of Geography, University of Sydney, Australia.

Selma Tozanli, PhD, is Senior Researcher at the Centre International des Hautes Etudes Agronomiques Méditerranéennes–Institut Agronomique Méditerranéen de Montpellier, France.

Nicholas von Tunzelmann, PhD, is Professor of the Economics of Science and Technology and Director of Research at SPRU, University of Sussex, United Kingdom.

Cláudia Viegas is a PhD Candidate at the Department of Economics, University of São Paulo (USP), Brazil. She is a lecturer at the Faculty of Economics and Business Administration of USP and a researcher at PENSA–Food and Agribusiness Program.

Thomas Worth, PhD, is Senior Economist with the Risk Management Agency, U.S. Department of Agriculture.

Acknowledgments

The editor has benefited from the advice and the comments made by Amarjit Basra, Christine Bolling, Anita M. Benvignati, Rolf Hackmann, Marina Papanastassiou, Bill Pritchard, Prasada Reddy, Roberto Simonetti, Victoria Salim, Howard Shatz, Paz Estrella Tolentino, Rob van Tulder, and Nick von Tunzelmann. Any remaining errors are her own responsibility. The editor also gratefully acknowledges the support of the Institute of Economics and Geography (IEG)-National Research Council of Spain (CSIC).

Introduction

Ruth Rama

As the process of globalization of the economy advances, international business studies have been required to investigate new issues and to revitalize old ones (van Tulder, van den Berghe, and Muller, 2001). Both managers and scholars have inquired whether companies should go international; if firms from any country—even developing countries—could become multinational enterprises (MNEs); and whether high levels of internationalization actually help multinationals to perform better. In home and host countries of MNEs, the process of globalization itself has aroused both expectations and suspicions. Would the worldwide expansion of MNEs substitute for or complement international trade of goods? If MNEs decided to use their own international networks for exchanges, would intrafirm trade (as opposed to trade in markets) then become the norm? Are MNEs likely to take their R&D facilities to foreign countries, depriving their home base of a substantial source of knowledge and high-tech employment? Given the progress of technoglobalism, would national systems of innovation become superfluous (Archibugi and Michie, 1995; Cantwell, 1995)? How would new regionalization processes, such as the development of the North American Free Trade Agreement (NAFTA), the European single market, or MERCOSUR (Mercado Comun del Cono Sur [Southern Core Common Market]), affect the globalization of firms?

The elaboration of these discussions also includes the need to revisit methodological issues and models. Most research in international business is still at a highly aggregated level. However, a growing number of scholars believe that if we could produce more empirical research at the company level, we could more accurately represent and understand MNE behavior (Caves, 1998; Frankho, 1989). There seems to be a need to not only continue to research trends at an aggregate level but also complement these through more detailed and rigorous firm-level research.

Though an enormous number of writings have tried to respond to such imperatives, the part allotted to multinational agribusinesses (MAs) in the discussions has been limited. Multinational agribusinesses are multinational enterprises active in (1) food and drink processing or (2) upstream industries, such as agriculture, aquaculture, etc., or in both. According to the International Labour Organization, the top 100 food and drink agroprocessing enterprises controlled 40 percent of the world market as of January 2001.

In spite of the importance of multinational agribusinesses, what we know about them derives mostly from cross-sectional studies on MNEs. However, in such studies, MAs are often represented by small samples of giant U.S. companies, insufficiently illustrative of the different sizes and nationalities that define new entrants to the international market in the contemporary era. Moreover, according to Goodman (1997), the literature on globalization of agrofood systems is often "mesmerized by parallels" with other sectors and could insufficiently take into account the particularities of MAs.

This book places the focus on MAs in the context of these debates. Putting the emphasis on MAs, it studies—in addition to the historically giant food firms—companies of various sizes and nationalities. It also identifies particular factors (for instance, food retailing) that specifically affect MA behavior. Responding to experts' requests, some of the econometric analyses in this book are focused at the firm, and even at the within-group, level. Other chapters are case studies on firms coming from nontraditional exporters of agrofood foreign direct investment (FDI). The book also provides a wealth of statistical information on the world's 100 top MAs, which could encourage more research on such companies in the future. On the other hand, the general debate running in international business studies is not forgotten, and the book attempts to identify, in the light of the discussion on MNEs, some of the distinct traits of MAs.

The book begins by considering one of the most important structural changes within economic globalization to have occurred in recent years, namely, changing modes of entry for FDI. By the mid-1990s, mergers and acquisitions accounted for around 78 percent of FDI in the world (Chesnais and Simonetti, 2000) and became the main mode for entry in foreign markets (van Tulder, van den Berghe, and Muller, 2001). MAs are no exceptions to such trends. Based on unique statistical evidence covering the 1974 to 2000 period, Tozanli analyzes, in Chapter 1, recent directions toward globalization among the world's 100 largest MAs, with special emphasis on mergers and acquisitions, closings, divestitures, and concentration in core activities. Arguing that not all MAs expand geographically in the same way, she supplies indicators of geographical and sectoral globalization for the 100 largest MAs over the period from 1988 to 2000. She also investigates the variations, in that period, in the positions of top MAs in terms of both their geographical dispersion and sectoral concentration.

As shown in Chapter 1, no less important than modes of entry are the strategies of expansion of global firms. In Chapter 2, Anastassopoulos and Rama study the impact of geographical and industrial diversification on the growth rates of the world's 100 largest MAs. For international managers, a matter of concern is that MAs have grown more slowly and became less profitable by the later 1990s, though there have been exceptions. Why have some companies performed better? Those that have grown faster, the authors note, are smaller and capital-intensive companies that avoided diversification into retailing and nonfood businesses but spread, instead, into food-related technology, such as biotechnology or specialized services to farmers. Then the authors tackle one of the most debated questions in the literature on MNEs: Does a higher level of internationalization lead to better performance as held by international management theory (van Tulder, van den Berghe, and Muller, 2001) or, by contrast, to poorer performance, as maintained by some Penrosian theorists? Anastassopoulos and Rama distinguish, as stated, different facets of multinationality. What seems to positively influence the growth rate of sales among MAs is country spread, while other aspects of multinationality could even exert negative effects. Over the recent period, MAs actually multiplied their geographic markets, while at the same time checking the development of their networks in host countries and maintaining the same balance of foreign versus domestic activities.

Though food and drinks is usually considered a low-tech industry, another factor that influences the performance of MAs is technical innovation and, given the importance of marketing for food and drinks, good packaging design. In Chapter 3, Alfranca, Rama, and von Tunzelmann review the literature on both technical and design innovation in such companies and provide new evidence obtained from a large database of patents granted to major MAs. They argue that the world's production of food and food-related innovation is led by a small nucleus of innovative MAs chiefly rooted, in spite of increasing globalization of their R&D, in home laboratories. Such companies, they suggest, display unique innovative traits versus other MNEs, such as their activities in design and their marked multitechnology characteristics. Unlike other multinational sectors, dominant MAs are not dislodged by "new" inventors. Since knowledge is accumulative in the agrofood multinational industry, the current level of innovation in MAs depends on their past innovative history; it also responds to pressures from rival innovators in the same line of business.

While the three first chapters focus on aspects related to MAs based in a variety of home countries, the five chapters that follow analyze evidence on MAs in specific settings (the United States, France, Australia and New Zealand, and Southeast Asian countries). The accelerated process of internationalization in food and food-related industries has modified the geographical direction and strength of FDI flows. The United States, traditionally a major investor in the food and kindred industries of foreign countries, has also become a recipient for foreign-based MAs. After analyzing recent trends for FDI in agriculture-related industries, Pick and Worth study, in Chapter 4, the characteristics of U.S. affiliates of foreign MAs, most of which are from firms based in other developed nations. They then focus on outflows of FDI from the United States to the food and kindred industries of other countries, most of which are also developed nations. These trends confirm the importance of cross investments in the food and drink industry (Rama, 1992). Finally, the authors pose the classic question of why foreign direct investment takes place, in order to highlight similarities and differences between the motivations of MAs and MNEs in other sectors.

In business science, one of the most debated issues since the early 1970s has been the relationship between FDI and trade. Analyses published in the 1970s anticipated that regionalization processes, such as European integration, could stimulate a concentration of production in the home countries of MNEs, who would rather export their products than produce abroad (Dunning, 1997). In practice, this theory did not hold true; FDI and trade actually seem interconnected.

According to the United Nations Conference on Trade and Development (UNCTAD), two-thirds of world trade takes place on the basis of direct involvement of MNEs (Chesnais and Sailleau, 2000). In the case of MAs, the analysis of the relationship between FDI and international trade is especially relevant for countries such as the United States and France, which are large exporters of food and agricultural products as well as traditional exporters of agrofood FDI. The next two chapters tackle the question.

In Chapter 5, Barkley summarizes and extends previous literature on the relationship between FDI and trade by empirically analyzing the determinants of the exports of the U.S. food-processing and agricultural industries. His econometric model shows that when U.S.-based firms increase their foreign investments, U.S. exports of food and related industries also tend to increase, which suggests that FDI and exports are complementary in most cases. However, he identifies some exceptions. When U.S. firms increase their FDI in some large agricultural export countries (the European Union [EU], some South American nations, and Australia), U.S. food exports tend to fall since, logically enough, such companies seem to source their inputs in the host country. In other words, concerning some of the most important host countries for agrofood FDI, the study could confirm previous findings of case studies in that major U.S. processing firms tend to integrate with local supply sources (Goodman, 1997).

After the United States, France is the world's largest exporter of food, as well as the most important in the EU. In Chapter 6, Chevassus-Lozza, Gallezot, and Galliano study, with unique empirical evidence, intrafirm trade in French agribusiness. As noted by Goodman's (1997) review, most authors researching agrofood systems view intrafirm trade as the hallmark of company globalization, since this practice evidences the presence of a division of labor within the multinational network. Working at the within-group level, Chevassus-Lozza, Gallezot, and Galliano analyze differences between the determinants of foreign trade (i.e., exchanges in the market) and intrafirm trade. The latter tends to grow, they note, when MAs try to valorize their specific advantages, to protect themselves against competition, and to exploit economies of scale. MAs that opt for internal trade are often experienced companies operating in concentrated agrofood markets. The study points to two important differences between MAs and other MNEs. First, intrafirm trade is less important among the former. Second, MAs exchange finished products with their affiliates for variety; unlike other MNEs, they rarely attempt to integrate different phases of production across borders. Within the MA, therefore, perishable, bulky inputs determine specific characteristics of the international division of labor.

In recent years, another prominent debate on the MNE has concerned the importance of the home base for firm strategy (van Tulder, van den Berghe, and Muller, 2001). The next two chapters discuss cases of nontraditional exporters of FDI in food and related industries—both developed and developing countries—and the conditions under which, eventually, their agrofood firms have gone successfully international.

As stated, MAs seem to generate, though with some exceptions, extensive international food trade. Conversely, is a country's competitiveness in food trade a point of departure for its agrofood firms going international? In Chapter 7, Pritchard analyzes the path to internationalization of Australian and New Zealand food multinationals. Since the 1980s, he explains, some have viewed the pioneer liberalization of agrofood trade in both countries as the first step toward the emergence of powerful MAs located in these nations. Based on evidence provided by in-depth analyses of three major companies, however, the author notes a weak relationship between the agrofood competitiveness of such countries and the emergence of successful MAs. The export competitiveness of Australian and New Zealand firms could, actually, make them targets for takeovers from foreign MAs rather than promoting their productive internationalization.

Chapter 8 continues the analysis of nontraditional exporters of agrofood FDI. In that chapter, Burch and Goss revisit, with a perspective from Southeast Asia, a classical question already discussed, mostly with evidence for Western firms, in Chapter 4: Why does FDI take place? In this case, the authors inquire, "Why would companies located in capital-poor and labor-abundant economies undertake FDI?" In the light of the literature on third world MNEs and using case studies of two major companies based in Thailand and Indonesia, Burch and Goss identify distinctive factors of success in South-Asian MAs that could account for their growth and international development, even in the face of the 1997 Asian crisis. The authors thus provide a primarily affirmative response to the question of whether food firms from developing countries can become MAs. Finally, they conclude by identifying differences between Southeast Asian MAs and Western MAs and challenge a number of theoretical assumptions on MAs which—in their opinion—often take only Western MAs into account. The reader will actually find some differences in factors of success (for instance, the role of vertical integration with retailing) in this book itself (see Chapter 2).

Finally, Chapter 9 contrasts with Chapter 8 in that it studies the case of a developing country as a recipient for *inward* agrofood FDI. It is widely admitted that the entry of affiliates in a developing country could have effects, both positive and negative, that are substantially different from those taking place in developed host countries (Blomström and Kokko, 1996). The ongoing globalization debate has strengthened concerns that the effects for developing host countries could primarily be negative (van Tulder, van den Berghe, and Muller, 2001). By focusing on the impact of agrofood FDI in Brazil—a major producer and exporter of food in the developing world—Farina and Viegas point to some negative effects but mostly to positive ones. They maintain that new recent macroeconomic policies (liberalization and stabilization of the economy) and the expansion of the consumer market have attracted to Brazil many MAs who have chosen it as a headquarter for their MERCOSUR operations. Their mode of entry, mergers and acquisitions, had negative effects on concentration and denationalization of the Brazilian food and drink industry, especially among top firms. In spite of this, they note, prices of some foodstuffs have fallen and second-tier brand names have enjoyed good fortune. The enlargement of the internal market and the concomitant macroeconomic changes, Farina and Viegas admit, could have also played their part. Nevertheless, they argue, the expansion of foreign retailers and increased levels of competition brought by new entrants—many of whom are second-tier firms from other Latin American countries, or from Italy or Ireland—have been instrumental in making some processed foodstuffs more available to the population and, eventually, in supplying better-remunerated jobs in the food industry. The globalization process, the study suggests, embraces the complex interplay of a variety of forces: different types of MAs—large established firms and new entrants; foreign retailers; and policies that actively encourage inward FDI and food imports.

REFERENCES

Archibugi D. and Michie J. (1995). The globalisation of technology: A new taxonomy. *Cambridge Journal of Economics* 19: 121-140.

Blomström M. and Kokko A. (1996). The impact of foreign investment on host countries: A review of the empirical evidence. Policy Working Paper Number 1745. Washington, DC: World Bank.

Cantwell J. (1995). *Innovation in a global world: Globalisation does not kill the need for national policies.* London: The Dryden Press.

Caves R. E. (1998). Research in international business: Problems and prospects. *Journal of International Business Studies* 29(1): 5-19.

Chesnais F. and Sailleau A. (2000). Foreign direct investment and European trade. In F. Chesnais, G. Ietto-Gillies, and R. Simonetti (Eds.), *European integration and global corporate strategies* (pp. 25-51). London: Routledge.

Chesnais F. and Simonetti R. (2000). Globalization, foreign direct investment and innovation. In F. Chesnais, G. Ietto-Gillies, and R. Simonetti (Eds.), *European integration and global corporate strategies* (pp. 3-24). London: Routledge.

Dunning J. H. (1997). The European internal market programme and inbound FDI. *Journal of Common Market Studies* 35(1): 1-30.

Frankho L. G. (1989). Global corporate competition: Who's winning, who's losing, and the R&D factor as one reason why. *Strategic Management Journal* 10: 449-474.

Goodman D. (1997). World-scale processes and agro-food systems: Critique and research needs. *Review of International Political Economy* 4(4): 663-687.

Rama R. (1992). *Investing in food.* Paris, France: OECD Development Centre Studies.

van Tulder R., van den Berghe D., and Muller A. (2001). *The world's largest firms and internationalization.* Rotterdamn, the Netherlands: Rotterdam School of Management/Erasmus University.

Chapter 1

The Rise of Global Enterprises in the World's Food Chain

Selma Tozanli

INTRODUCTION

International competitive dynamics, a direct result of the globalization process, relies more and more on total integration of companies into global business networks. Seeking efficiency in continuously enlarging markets becomes a condition for worldwide competitiveness. A successful and durable enterprise "operates with resolute constancy as if the entire world or major regions of it were a single entity, selling the same things in the same way everywhere" (Levitt, 1983). Only those enterprises that manage to organize their international business activities within global and networking structures can hope to attain durable worldwide competitiveness.

Consumers' needs are currently homogenized and result in internationally standardized, uniform products pushing the markets toward a global commonality (Levitt, 1983). Multinational food-processing enterprises (MNEs) not only have to deal continuously with the perishable nature of food but also have to take into consideration important social and cultural aspects of eating habits, individual preferences, and tastes specific to a given socioeconomic environ-

I address my grateful thanks to all the members of the ERFI-SIF and UMR MOISA Research Teams, especially to Florence Palpacuer, Amélie Seignour, Fatiha Fort and Roland Pérez, Jean-Louis Rastoin, and Gérard Ghersi for the precious advice. I also thank Mrs. Goodman-Stephens for her invaluable linguistic assistance.

1

ment. These aspects evolve in a specific socioeconomic environment which calls for a particular globalization pattern. An analysis of this globalization pattern of the world's agrofood chain is the main subject of the chapter.

First, I will analyze the essential drivers that conduct food-processing multinational enterprises (food MNEs) toward building stable market positions on a worldwide basis. Then I will show the orientation and scope of their recent geographical expansion and their sectoral reconfiguration resulting from their refocusing strategies on "core" business activities. These preliminary indicators allow us to trace the changing composition of the world's food oligopoly represented by the top-100 leading food-processing MNEs. I base the empirical analysis on data from our databank, Agrodata, developed at Centre International des Hautes Etudes Agronomiques Méditerranéennes–Institut Agronomique Méditerranéen de Montpellier (CIHEAM-IAMM), where qualitative and quantitative information on the world's largest food-processing multinational firms has been compiled since the 1970s. According to Agrodata, an MNE is an enterprise that has at least one food-processing plant located outside the home country of the parent company. Another selection criterion is that the size of the MNE, measured by its worldwide agrofood sales, must exceed US$1 billion (Rastoin et al., 1998). In addition to highlighting qualitative information on origin, historical development, strategic moves, and the main business activities of these firms,[1] Agrodata also compiles information on the number, the nature of the business activity, the host country, and the percentage of control of their consolidated subsidiaries. Consolidated financial data, as reported by food MNEs, could be used to calculate growth trends and economic and financial ratios over several decades. Furthermore, Agrodata has, since the mid-1980s, compiled data on merger and acquisition (M&A) operations, partnerships, joint ventures, sell-offs, divestitures, spin-offs, and plant closures realized by food MNEs. These data are not exhaustive and are limited by the nature and the range of documentary resources.[2] A periodical selection made among the firms within the Agrodata database has resulted in a world ranking of the top-100 food MNEs.

A DICHOTOMOUS GLOBALIZATION PROCESS

Recent changes in the composition of the top 100 are deeply rooted in the way food MNEs respond to new international competitiveness. In fact, a new balance of power between mature industrialized economies and emerging growth-oriented economies seems to determine market strategies and to give growth orientation to food MNEs. As mentioned earlier, globalized urban lifestyles, the increasing proportion of economically active females, the worldwide acculturation process, the increasing number of people living on their own, and childless households are the recent *globalizing drivers* that bring the populations of megalopolises throughout the world closer together and call for processed, user-friendly food products.

Aided by the internationalization process of large food retailers, the top 100 have eroded mature Western markets which now exhibit stagnant demographic trends and aging population features. The top 100 are gradually moving toward emerging economies that exhibit both positive demographic trends and increasing consumer purchasing power. However, this move is not equal for all MNEs in the top 100, nor is it evenly distributed from one continent to another. Furthermore, as shown in Table 1.3, the importance of the Triad (North America, Japan, and Western Europe) as the pulling pole of MNE investments continues (Markusen, 1998) to prevail despite the recent geographical expansion of the top 100. This biases somewhat the growth trends of the MNEs (see Figure 1.1).

Thus, in their move toward worldwide expansion, the top 100 reveal a dual growth trend. They concentrate their resources in a small number of main business activities that ensure them leader or challenger positions in international markets. While globalizing steadily in mature Western markets, they adopt a "glocal" and/or "multidomestic" strategy in emerging markets with high growth potential (Porter, 1986; Mucchielli, 1998). I will try to explain this dichotomous, yet not contradictory, evolutionary path. My arguments are based on empirical data from the database concerning all of the subsidiaries in the top 100 in 1988 and in 1999-2000 and the total number of restructuring operations realized by the same top 100 between January 1, 1987 and January 31, 2002 (see Tables 1.1 and 1.2).

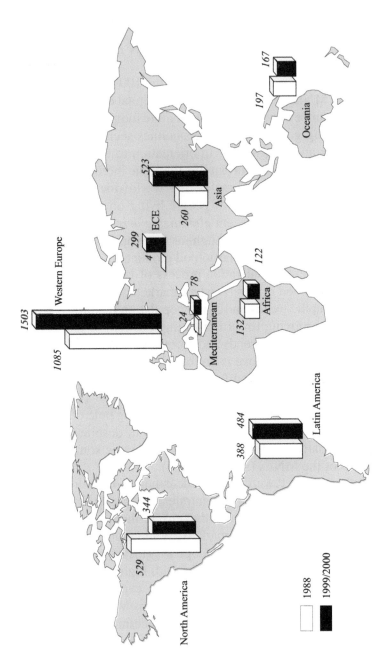

FIGURE 1.1. Distribution of the foreign subsidiaries of the top-100 MNEs operating in the agrofood sector according to host zones in 1988 and in 1999-2000. (*Source:* Agrodata database, CIHEAM-IAMM, Montpellier.)

North America
529
344

Latin America
388 484

Western Europe
1503
1085

ECE
299
4

Mediterranean
24 78

Africa
122
132

Asia
523
260

Oceania
197 167

1988
1999/2000

TABLE 1.1. Percentage distribution of acquisitions and majority takeovers,[a] January 1, 1987, to January 31, 2002.

Home zone of the top-100 MNEs	Africa	Latin America	North America[b]	Japan	Asia	ECE[d]	Mediter-ranean[c]	Oceania	Western Europe	Home country of the parent company
Before 1992[d]										
Western Europe	0.6	2.8	20.5	0.1	0.4	1.0	0.1	2.8	39.0	32.5
United States	0.3	5.3	8.3	0.3	0.3	0.6	–	1.2	24.8	59.0
Japan	–	–	38.5	–	7.7	–	–	19.2	26.9	7.7
Rest of the world	–	3.0	19.7	–	3.0	1.5	–	7.6	31.8	33.3
Top 100	0.4	3.5	17.1	0.2	0.7	0.9	0.1	3.0	34.0	40.0
After 1992										
Western Europe	1.6	9.5	12.7	0.4	5.3	12.9	2.4	1.8	30.1	23.1
United States	0.6	11.7	3.6	0.2	5.3	7.2	2.1	3.8	26.7	38.0
Japan	–	–	36.0	–	8.0	–	–	8.0	24.0	24.0
Rest of the world	2.3	10.5	18.6	–	12.8	11.6	–	3.5	12.8	27.9
Top 100	1.3	10.0	10.7	0.3	5.8	11.0	2.1	3.0	28.1	27.7

Source: Agrodata database, CIHEAM-IAMM, Montpellier.

[a]"Majority takeover" is defined as an acquisition transaction of 50 percent or more of the capital of the target firm.
[b]North America, as defined in Agrodata, comprises the United States, Canada, the Bahamas, and Bermuda.
[c]The Mediterranean zone comprises, according to the definition by Pr. Louis Malassis (1971) and based on the work of F. Braudel (1985), the Balkan countries, Turkey, Iran, Iraq, Syria, Lebanon, Jordan, Israel, Egypt, Saudi Arabia, United Arab Emirates, Yemen, Kuwait, Bahrain, Sudan, Libya, Tunisia, Algeria, Morocco, Gibraltar, Malta, and Cyprus. Member states of the EU are included in the Western European zone.
[d]1992 is considered to be the cutoff point in order to lessen the biasing effect of the insaturation of the European single market.

TABLE 1.2. Percentage distribution of mergers, minority takeovers,[a] partnerships, license and franchising agreements, January 1, 1987, to January 31, 2002.

Home zone of the top-100 MNEs	Africa	Latin America	North America	Japan	Asia	ECE	Mediter-ranean	Oceania	Western Europe	Home country of the parent company
Before 1992										
Western Europe	2.1	2.8	8.3	4.0	8.0	4.6	1.2	1.8	37.8	29.1
United States	–	3.8	1.9	3.8	15.1	17.0	3.8	1.9	26.9	26.4
Japan	–	1.5	13.4	–	20.9	1.5	–	7.5	28.4	26.9
Rest of the world	–	2.7	13.5	2.7	2.7	–	–	8.1	37.8	32.4
Top 100	1.3	2.8	8.0	3.4	10.6	6.3	1.5	3.0	34.4	28.5
After 1992										
Western Europe	1.0	6.2	7.1	2.1	16.4	15.2	5.2	2.4	25.0	18.8
United States	3.3	12.7	5.8	1.9	7.3	9.3	5.3	2.0	25.3	30.0
Japan	–	–	7.0	–	51.2	7.0	–	–	18.6	16.3
Rest of the world	–	12.0	12.0	–	16.0	12.0	–	4.0	16.0	24.0
Top 100	1.4	7.5	5.8	1.9	16.6	13.2	4.7	2.2	24.3	21.5

Source: Agrodata database, CIHEAM-IAMM, Montpellier.

[a]"Minority takeover" is defined as an acquisition transaction of less than 50 percent of the capital of the target firm.

Globalization of Food MNEs in Mature Western Markets

The downsizing and streamlining wave that hit the U.S. food-processing sector in the late 1970s and throughout the 1980s has been striking the European market since the early 1990s. Many of the European food giants applied successive restructuring programs to obtain scale and scope economies in their industrial production and organizational management. One of the most important reasons that led to the adoption of such programs has been explained by top managers as the need to coordinate the dispersed activities of their corporation around the company's core activities, organized in a cross-divisional manner so as to permit a better flow of information and decisions (Figure 1.2).

The dispersal of activities was a natural outcome of the decade of mergers and takeovers, in which each subsidiary continued to manage its activity as an independent unit. Consequently, giant food

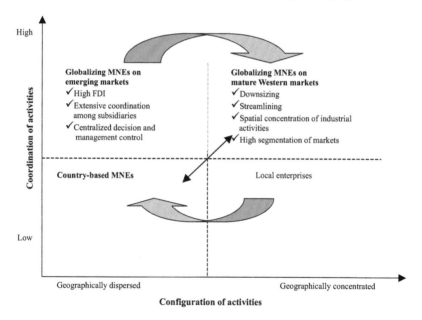

FIGURE 1.2. Globalization strategies of food-processing MNEs. (*Source:* Tozanli, UMR MOISA, CIHEAM-IAMM, Montpellier, 1999, based on the schema of Porter, 1986, p. 28.)

MNEs, especially the multiproduct multinationals,[3] resembled federations gathering within their corporate organizations, a multitude of independent small and medium-sized enterprises (SMEs). Duplication of a number of functions by these hundreds of subsidiaries and dozens of divisions that composed these giant multiproduct multinationals (e.g., Unilever, Nestlé, Sara Lee Corp.) reduced the firms' profitability and economic efficiency. MNEs were, however, reluctant to adopt streamlining and downsizing programs because the European single market was shrinking, and transportation costs were decreasing at the same time that, in many of their processing plants, production costs were increasing with overcapacities. Consequently, they decided to base their production design on a continental structure whereby huge industrial plants supply different national markets simultaneously while remaining in constant touch with retailers' demands. In order to increase their production capacities and realize large-scale economies with flexible production plans, they shut down many outdated plants to automate and modernize others.

When we look at the number of plant closures that took place between January 1, 1987, and January 31, 2002 (Table 1.3), we see that 78 percent occurred in Western European countries, 8 percent in North America, and 5 percent in Japan.[4]

On the other hand, MNEs opted for high value added, environmentally friendly "niche" products in order to meet the demands of an ever-more volatile European consumer. An aging demographic structure, constantly increasing health concerns related to food consumption, more widespread adoption of urban lifestyles, and an increasing number of people living on their own or in childless households are the main limiting factors responsible for the shrinking market in industrialized Western countries. Furthermore, we see a smooth shift of consumer choice toward unbranded, organic, and/or genuine local products, particularly in Western European countries, which increases the threat to the future development of large food-processing MNEs.

Many of the leading MNEs sell off the divisions or subsidiaries that do not fall within their core activities. All those activities that do not correspond to a targeted profitability or return on investment ratio calculated on a mid-term or long-term basis, and/or that do not show up in worldwide leading positions, are considered outlying activities and are sold off as soon as possible (Pérez and Palpacuer, 2002). We

TABLE 1.3. Percentage distribution of plant closures, January 1, 1987, to January 31, 2002.

Home country (zone) of MNEs	Latin America	North America	Asia	ECE	Oceania	Western Europe	Home country of the parent company
Canada	–	–	–	–	–	50.0	50.0
United States	16.7	4.7	–	–	–	37.5	41.7
Japan	–	–	–	–	–	–	100.0
Netherlands	3.3	1.0	–	–	–	60.0	26.7
Norway	–	–	–	–	–	–	100.0
Switzerland	20.0	–	–	–	–	70.0	100.0
Germany	–	–	–	–	50.0	–	50.0
Belgium	–	–	–	–	–	100.0	–
France	–	–	–	6.3	–	31.3	62.5
United Kingdom	–	–	–	–	–	11.5	88.5
Ireland	–	–	–	–	–	100.0	–
Italy	–	–	–	–	–	100.0	–
Western Europe	3.3	3.3	–	1.1	1.1	42.2	48.9
Top 100	5.8	3.3	–	0.8	0.8	39.2	50.0

Source: Agrodata database, CIHEAM-IAMM, Montpellier.

can once more refer to the empirical data concerning mergers and acquisitions for the same period (January 1, 1987 to January 31, 2002) even if these data are not exhaustive. Out of 1,043 operations that were recorded as divestitures realized by the top-100 food MNEs (selling off of subsidiaries and of minority or majority shares), 66 percent took place in Western European countries and 29 percent in North America. Similarly, one fifth of the seventy-six mergers concerning the subsidiaries of top-ranking MNEs were realized in North America, while 61 percent took place in Western European countries (Table 1.4).

On a wider scale, during the 1990s, the top-ranking food MNEs rushed to sell off their agricultural activities to pharmaceutical or chemical companies. Newcomers to the world's food oligopoly are mostly specialized in capital-intensive, high value added food-processing branches such as probiotics or other functional foods, food ingredients, nutroceuticals, and clinical food. New entrants in upstream activities specialize mostly in poultry, aquaculture, and pisciculture. They are highly specialized, capital intensive, and highly technology oriented, just the opposite of those companies that dealt with upstream activities in the 1970s. The latter were, on the whole, large subsidiaries of highly diversified industrial conglomerates or large internationalized agricultural cooperatives.

Most of the large food MNEs also dropped marginal products that could not be marketed beyond regional or national markets. They instead concentrated their financial, human, and capital resources on a small number of core activities and worldwide famous brands, which could be marketed on a global scale (see Box 1.1). As a result of this centering movement, we observed the comeback of "single-product" MNEs such as Wm. Wrigley Junior and Mc Cormick and Co. in our world ranking along side large "packaged consumer product" conglomerates that refocus their worldwide businesses around their main "firm-specific activities" (Rugman, 1998).

Geographical Expansion Beyond the Triad

Food MNEs base their presence beyond their national boundaries on their well-known brands. In fact, they have a dispersed multi-domestic configuration concerning their production pattern in most of the developing regions. External growth via M&A operations and

TABLE 1.4. Percentage distribution of sell-offs, divestitures, and spin-offs, January 1, 1987, to January 31, 2002.

Home zone of the top-100 MNEs	Africa	Latin America	North America	Japan	Asia	ECE	Mediter- ranean	Oceania	Western Europe	Home country of the parent company
Before 1992										
Western Europe	0.3	0.6	20.0	0.6	0.9	–	–	1.5	19.4	56.8
United States	1.3	2.5	5.7	–	0.6	–	–	1.9	28.3	59.7
Japan	–	–	–	–	–	–	–	–	–	–
Rest of the world	2.0	–	31.4	3.9	7.8	–	–	6.0	11.8	43.1
Top 100	0.7	1.1	16.9	0.7	1.5	–	–	2.0	21.3	56.4
After 1992										
Western Europe	1.4	2.3	11.1	0.8	2.5	1.8	0.2	1.6	38.5	40.6
United States	1.0	8.6	2.4	0.5	3.3	4.8	0.5	2.9	23.4	53.1
Japan	–	–	33.3	–	–	–	–	66.7	–	–
Rest of the world	–	–	12.5	10.4	6.5	–	–	4.2	53.3	10.4
Top 100	1.2	3.9	8.8	1.3	2.9	2.5	0.3	2.4	33.8	42.0

Source: Agrodata database, CIHEAM-IAMM, Montpellier.

BOX 1.1. Unilever's Divestitures

One of the best examples of divestitures is Unilever, which sold off its chemical division, considered not to be part of its core business activities, to I.C. Industries which subsequently strengthened its lead position in the world market thanks to that very acquisition. Meanwhile, Unilever also got rid of its industrial services division "DiverseyLever" to refocus on its core activities. Following a horizontal concentration process, from the beginning of the 1990s, it placed all its effort into the ice cream business and into highly processed foods. Another striking example is Groupe Danone, which restructured its industrial production and divisional organization over only a few years starting in 1994 when it announced its aim to globalize. The French MNE organized its industrial activities around three core business activities in which it had lead or challenger positions, namely fresh dairy products (worldwide leader), bottled water (worldwide challenger behind Nestlé) and biscuits (worldwide challenger behind PepsiCo).

This solid position demanded sacrifices both from the company's top management and from its stakeholders, mainly the French government and its European salaried employees. Groupe Danone first went through a slimming phase, selling off its grocery division, subsidiary by subsidiary, and by closing down some of its European plants specialized in processing fresh dairy products. Soon after, it got rid of its beer processing division Kronenbourg which was considered by top management as not promising in terms of worldwide development. Its biscuit processing plants went through a streamlining and downsizing program which caused social outcry especially in Western regions of France. As a result of these modifications, Groupe Danone has emerged as one of the world's leading global food processing MNEs.

---

strategic alliances help them to avoid market risks as well as sunk costs arising from greenfield investments,[5] while the local lead firms, by becoming MNE partners, bring along 30 to 50 percent of the local market shares.

In recent years, this trend has gained commonality, as external growth has been adopted by the majority of food MNEs. Transactions are generally based on the specific assets of both partners. A food MNE brings along its global brand name, worldwide reputation, and world distribution network. If necessary, it also supplies technological assistance, human capital, or capital improvement investments. Its involvement is best illustrated by the management control that it exerts over the functioning of its local partner, controlling product and information flow. This partnership facilitates the access of the local lead firm to international markets by upgrading its production standards. Last but not least, it also pours in cash that all growth-oriented firms seriously need.

In return, the local lead company provides its local distribution channels, its experiential knowledge of its own local market, as well as its know-how concerning local dishes and industrial processing techniques. These local market-specific assets help the food MNE to keep pace with local market conditions, which can otherwise be particularly difficult to deal with. Local business habits, often the main obstacle to the success of an MNE, can be successfully managed thanks to local partners.

But do all food-processing MNEs expand geographically in the same way? Can we talk about a general globalization trend among the major players in the world's food oligopoly? In order to answer these questions, I tried to measure the degree of globalization of leading food MNEs. I am aware that the globalization process is a budding phenomenon in the world's food-processing industry and that global firms, as defined in general literature on economics and strategic management, are hardly present among the food-processing enterprises (Porter, 1986). I suggest, therefore, that a globalizing food-processing MNE be defined as an enterprise that expands its food-processing activities in overseas markets.

We can measure the *geographical dispersion intensity* (GDI) of an MNE by dividing the number of host zones receiving its foreign subsidiaries with food-processing activities, by the total number of geographical zones as defined by Agrodata.[6] (For other measures see Chapter 2.) Second, we measure the *internationalization intensity* (IZI) by dividing the sum of its foreign subsidiaries by the sum of all of its subsidiaries,[7] including those located in the home zone of the parent company. Thus we obtain a *geographical globalization indicator* (GGI) by multiplying the geographical expansion intensity by the internationalization intensity.[8]

The geographical globalization indicator helps to extenuate the biases that can arise from the significant presence of an MNE in its home zone. For example, a Western European food MNE that has a great number of foreign subsidiaries in a single European market will be considered less globalized than another MNE that has a smaller number of foreign subsidiaries in the European Union (EU) and a certain number of subsidiaries more evenly dispersed over the eight geographical zones. The ideal case being, of course, a food MNE with a large number of foreign subsidiaries evenly spread over the seven other geographical zones beyond the home zone of the parent company.

Tables A1.1, A1.2, and A1.3 show the descending ranking of the leading food MNEs by their indicator of geographical globalization as well as the evolution that this process underwent over the period 1988 to 1999-2000. A majority of the food MNEs showed a tremendous ascending trend, especially with regard to geographical expansion intensity. The already highly internationalized multiproduct MNEs underwent the most significant change during the 1990s. If we aggregate these MNEs by their zone of origin, we see that MNEs originating from Western Europe seem to be much more dynamic than their U.S. or Japanese counterparts with regard to *geographical globalization intensity.*

Aside from those food giants originating from the United States that have already highly dispersed their business activities worldwide, most of the American food MNEs, even the best-known multinationals such as General Mills, Kellogg's, or Campbell Soup, have a less dynamic growth pattern in their geographical globalization process. The cultural proximity of the Western European or Pacific Rim markets, or the geographical proximity of some Latin American countries, secure the position of U.S. MNEs, while the high growth potential of emerging Asian, Mediterranean, or ECE markets seems to exert a limited pull factor for their foreign investments. Japanese MNEs also seem to be less attracted than the large, mature Western markets by the highly dynamic but less stable markets of emerging countries.

In addition to the significant constraints drawn up by the large Western European food retailers, the narrowness of the home market, stagnant and even decreasing demographic trends, and the aging populations of Western European countries are among the most important push factors exerting major pressure on Western European food MNEs. As mentioned, besides the already well-established globalizing multiproduct MNEs, we see the arrival of relatively medium-sized but highly dynamic food MNEs from small European countries.

Could these newcomers become the new major players that will reshape the world's food-processing landscape in the next decade, or will they be taken over and swallowed up by the huge multiproduct multinationals as has been the case over recent years? It is quite hard to predict.

SECTORAL GLOBALIZATION:
REFOCUSING ON CORE BUSINESS ACTIVITIES

Enterprises have limited resources, and the quest for sustainable international competitiveness leads them to use these in an ever more rational manner. A business strategy aiming to achieve economies of scale in order to keep high sales volume with a minimum cost structure on the one hand, and to realize economies of scope by sectoral diversification on the other, no longer responds to the efficiency seeking of the enterprise. Furthermore, the increasing pace and intensity of the internationalization process pushes these enterprises to be more economical in terms of their resource endowment. Motivated by the need to allocate their resources toward the production, promotion, and distribution of their most standardized, global-branded products, they look for synergies in larger markets (Foss and Iversen, 1997). Accordingly, they concentrate their capital as well as their human and financial resources on what they call "core" business activities and abandon those activities that appear peripheral (Palpacuer, 1999). Information flow, raw material and intermediary consumption procurement, management control, accounting principles, advertising, and R&D activities all become transversal on a very large basis in order to cover the totality of the subsidiaries dispersed worldwide. Therefore, synergies can help the parent company by ensuring the best use of these precious resources while maintaining an information flow that allows for rapid identification of any disfunctioning units.

Food MNEs aim to enlarge their worldwide market shares by horizontal concentration through establishing local partnerships. Some of these lead MNEs that opted to build global brand names provide some estimates of their world market shares (Aaker and Joachimsthaler, 1999). For example, Danone, with its global brand for yogurts and dairy desserts, owns approximately 10 percent of the world market shares. It is the world challenger behind Nestlé in the bottled drinking water market. It is presently targeting a leading position in the world's biscuit market. Similarly, PepsiCo, the challenger to the world leader Coca-Cola which has 51 percent of the volume of carbonated soft drinks produced worldwide, has captured 21 percent of the world industry. Concerning the world's salted snack food industry, the world brand Frito-Lay covers 40 percent of world production and thus pushes PepsiCo simultaneously to the world's number one position in this domain.[9]

All MNEs do not refocus on core business activities in the same way. Some of them, mostly those originating in Japan, still have a dual growth strategy. They expand beyond their national boundaries or home zone in only one or two business activities while continuing to keep a multiproduct/multidivision structure in their home markets. Their internationalization generally concerns the Triad markets only. On the other hand, most U.S.-based MNEs, which have been undergoing a centralization process for more than a decade (see Table A1.8), with some exceptions such as ConAgra, have decreased their number of business units.

Over the past five years, the evolutionary trend concerning huge mergers or takeovers shows that most of the MNEs absorbed by other giants are country-based MNEs with limited geographical expansion. This is also true for some British MNEs,[10] which were either bought by other firms or were left behind the top 100 due to the small growth observed in their overall agrofood sales. On the contrary, other Western European MNEs seem to challenge the globalization process not only by geographical expansion but also by sectoral reorganization of their businesses.[11]

The search for synergies between different business units, forming the internal structure of food MNEs, is accomplished essentially via restructuring operations. We observe a certain move-out trend from some sectors. Meanwhile, other sectors such as functional foods, diet foods, convenience meals, and nutraceuticals seem to exert a considerable pull effect and result in a great number of mergers and takeovers.

Data presented in Tables A1.1, A1.2, and A1.3 show that globalizing MNEs develop those business activities that can easily be duplicated and standardized in emerging markets: dairy products, biscuits and baking products, beer and spirits, and confectionery. Their worldwide reputation helps them increase the market shares that their partner enterprises or local subsidiaries already possess. With regard to emerging markets, dairy processing is an important domain for acquisitions and majority takeovers in Africa (17 percent), the Mediterranean (17 percent), the Eastern and Central European countries (ECE) (17 percent), and Latin America (14 percent). Alcoholic beverages and beer are highly attractive sectors for restructuring operations in ECE (30 percent), Asia (16 percent), and Africa (15 percent), while soft drinks are pull factors concerning FDI mostly in Africa (12 percent), Mediterranean countries (11 percent), and ECE (7 percent). Baking and pasta products, biscuits, and confectionery and chocolates are also attractive sectors for M&A operations realized in emerging markets. Meanwhile, among the total restructuring operations involving acquisitions and majority takeovers carried out in the Triad,[12] one-third were realized in dairy processing, 30 percent in high value added food processing, more than 11 percent in ready-to-eat meals and snacks, and nearly 6 percent in nutraceuticals, microbiotic products, and proteins.[13]

So, in light of this empirical evidence drawn from Agrodata, it can be argued that today the multinationality of a food-processing enterprise should be defined by *ownership advantages* of the firm, which are dependent upon intangible specific assets such as global brand name, MNE reputation, and management skills beyond technological and innovative specificities (Narula and Dunning, 1999). *Location advantages* offered by emerging foreign markets because of their high-growth potential are closely linked to these ownership advantages. *Internalization advantage* is losing its significance in the internationalization of food-processing enterprises, as the majority of food produced has a bulk character, i.e., it can easily be imitated and duplicated anywhere in the world by any enterprise that can master international standards. Instead, licensing and franchising as well as other partnership patterns are becoming common use even if the takeovers and mergers still keep their supremacy. Meanwhile, those MNEs that are establishing worldwide information and communication networks (Danone, Nestlé, Unilever, Sara Lee, HJ Heinz, etc.) prefer, when they can, to control the majority stake of their partners because of the growing strategic feature of knowledge and information within the organization system of the enterprise.

Those MNEs that do not manage to establish this three-bond advantage system (Narula and Dunning, 1999; Markusen, 1998) on a global basis do not acquire the necessary criteria for international competitiveness. Large industrial conglomerates that did not take advantage of their food-processing arms because they were not sufficiently global, thus not profitable, sold them during the 1990s. As a result, we have seen South African giant Barlow Rand, the U.S.

chemical giants WR Grace, Monsanto, Occidental Petroleum, British Petroleum (BP), and Swedish Volvo retire from food processing.

I based my arguments concerning the different refocusing programs used by leading food MNEs on data concerning all the subsidiaries in the top 100 for 1988 and 1999-2000 (Table A1.8). I counted the number of total business activities[14] in which each MNE was operating in those years both in its home and host zones. I adopted the notion of home zone rather than the home country of the multinational enterprise in order to take into account the changing structure of its natural markets. Therefore, the home zone of U.S. MNEs comprises all North American markets; the home zone of Western European MNEs is made up of the European single market as well as the rest of the European countries (Switzerland, Norway, Iceland); and for Japanese MNEs, the home zone includes the Asian market. This distinction helps us to see if our top MNEs really expand overseas. In my opinion, it is a better measurement of their globalizing feature, as it lessens the biasing effects of geographical and cultural proximities.

I obtained a *sectoral globalization indicator* (SGI) by dividing the number of business activities in which a food MNE operates within its host zone(s) by the number of its business activities in its home zone.[15] This indicator helps us to see if an MNE is a country-based ethnocentric MNE, a multidomestic polycentric MNE, or a globalizing geocentric MNE, according to Perlmutter's classification (Mucchielli, 1998). If an MNE has more business activities in its home zone than overseas, it is considered to be ethnocentric. A globalizing, geocentric multinational is one that has the same number of business activities in its home zone as in its host zones. A polycentric MNE is a multinational with a limited number of activities spread over the world and with a large activity base in its home zone.[16] Multiproduct giants such as Unilever, Philip Morris, and to a lesser extent Nestlé, considered to be polycentric multinationals, are undergoing important changes in order to decrease the number of their business activities both at home and in host zones.[17]

The global competitive dynamics of a geocentric multinational depend, according to this hypothesis, on the synergies that it is able to create between its domestic and international business activities, thereby better coordinating its different activities developed on a worldwide basis through vertical, centralized coordination. Thus, with regard to its economic and managerial efficiency seeking, the factor of achievement is the attainment of optimum organizational consistency through vertical, centralized coordination rather than the total number of its business units.

The globalization process is therefore brought about by an interaction between geographical globalization intensity (GDI index) and sectoral globalization (SGI index). Figures 1.3 and 1.4 show the junction of these two indexes on a matrix as well as the evolutionary paths of the top 100 during the last decade (1988 and 1999-2000). It is difficult to comment briefly on the evolution of a large population of multinational enterprises, as each has its own evolutionary path. However, we can say that one outcome of this evolution is that the average geographical globalization index increased from 20 to 42 percent. This shows that the majority of the top 100 tried to intensify their internationalization and geographical expansion beyond their home zone over that ten-year period. Regarding the average sectoral globalization index, the change was from 39 percent in 1988 to 52 percent in 1999-2000.

We cannot classify these MNEs as outright winners or losers because the world's competitive landscape changes so rapidly. Nevertheless, we can say that most of the multiproduct multinational giants mentioned here underwent an important refocusing on "core" activities in which they increased their presence overseas while decreasing the number of subsidiaries in their home zone. However, their huge size seems to be an important handicap to rapid change. Other smaller multiproduct multinationals such as Danone, Pernod Ricard, and Cadbury Schweppes managed to globalize more rapidly because of better flexibility.

MNEs dealing with commodities and primary processing of agricultural products[18] disappeared from the world ranking, because they were not sufficiently internationalized to compete with high value processing multinationals. Some of these MNEs such as ConAgra and ADM (Archer Daniels Midland) evolved toward high-value processing by adopting a strategy of differentiation. Commodity-based MNEs such as Sudzucker, ABF, The Kerry Group, Glanbia, Dean Foods, and IBP seem to have stayed within the danger zone, threatened by market pressures. Japanese MNEs, which underwent an evolutionary path counter to the general trend, increased their home zone business activities and their country-based subsidiaries. This may lead to an eventual loss of flexibility and market reactivity and could

FIGURE 1.3. Positioning of the world's top-100 MNEs according to their geographical and sectoral globalization indexes in 1988. (*Source:* Agrodata database, CIHEAM-IAMM, Montpellier.)

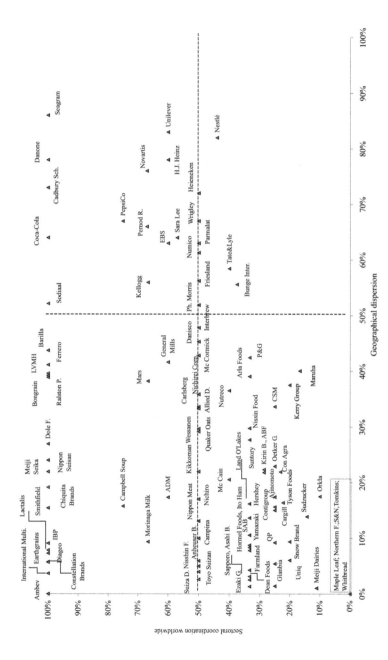

FIGURE 1.4. Positioning of the world's top-100 MNEs according to their geographical and sectoral globalization indicators in 2000. (*Source:* Agrodata database, CHIEAM-IAMM, Montpellier.)

21

threaten the independence of Japanese multinationals, as discussed in the next section.

Harmonization and coordination of its activities on a worldwide level lead an enterprise toward globalization. Decisions taken no longer concern only the home country or one single region. On the contrary, its entire activity is made up of standardized, interrelated, and coordinated subsidiaries operating according to international benchmarks defined by the parent company. Therefore, the supporting activities of the MNE such as procurement, technological development, and human resource management are globalized. This strategic move results in considerable improvement in the cost structure of the MNE (Porter, 1986) as has been the case for Nestlé, Groupe Danone, Unilever, Sara Lee, and HJ Heinz.

THE CHANGING STRUCTURE OF THE WORLD'S FOOD OLIGOPOLY

Today, as in the past, the world ranking of food MNEs shows the supremacy of the Triad, even though the composition within the Triad itself has changed. As the competitive dynamics change, the globalization process becomes one of the essential driving forces. Those enterprises that fall behind the requirements of this international competitiveness edge simply disappear.

Furthermore, food-processing capital, measured in our calculations as the overall agrofood sales of the leading food MNEs worldwide, is concentrated more and more in the hands of the largest multinationals, while the concentration curve smoothens after the fiftieth enterprise (see Figure 1.5). Also, environmental constraints, essentially pressure exerted by large retailers on the selling price, blocked any increase concerning the aggregated overall sales of the top 100 during the 1990s. Increasingly, newcomers tend to be moderately sized compared to the giants making up the largest ten MNEs.

One of the most striking changes is the decreasing number of U.S.-based MNEs (see Figure 1.6). The consolidation movements of this huge, processed food market by successive concentration waves have reconfigured the composition of the U.S. food oligopoly. As I have argued all along, the fact that some of the U.S. MNEs do not actually respond to the requirements of this globalization process also appears to be an essential reason for their failure. Some of the huge takeovers

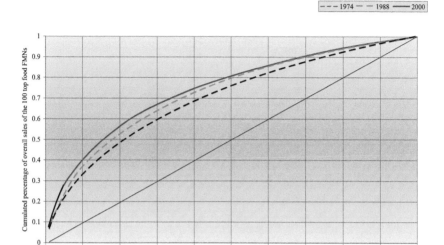

FIGURE 1.5. Concentration curve of the world's top-100 food MNEs. (*Source:* Agrodata database, CIHEAM-IAMM, Montpellier.)

signal the quest by U.S. food giants to enlarge their markets in size as well as in geographical scope.[19] As a result of this ongoing consolidation process, the average size of the U.S.-based food MNEs in the world ranking is increasing continuously. U.S. food multinationals had an average size of US$1.5 billion in 1974, and of US$8.1 billion in 1988. This tremendous increase continued during the decade that followed to reach an average size of US$12.3 billion in 2000. Takeovers during the 2001-2002 period seem to have accentuated this rising trend.

Another prominent change is the massive arrival of Japanese food MNEs among the world's top 100 (see Figure 1.6). In the 1970s and the 1980s, large Japanese food processing enterprises went multinational in order to avoid political constraints linked to changes in international trade rules. A clear example of this is the location of Japanese fishing companies in Latin American countries in order to bypass the 1974 International Law of the Sea that changed the scope of national territorial waters. Another example is linked to the liberalization of the Japanese food market induced by the General Agree-

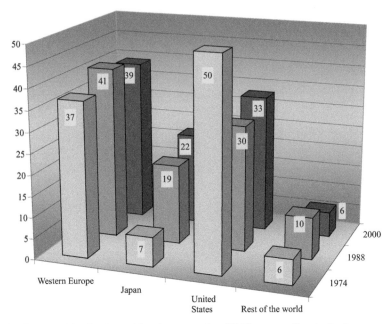

FIGURE 1.6. Distribution of food-processing MNEs according to their zones of origin. (*Source:* Agrodata database, UMR MOISA, CIHEAM-IAMM, Montpellier.)

ment on Tariffs and Trade (GATT) in the mid-1980s. Consequently, to keep their supremacy in their home market, Japanese meat processing companies (Nippon Meat Packers, Itoham Foods) bought up U.S. and Australian ranches, thus importing their own meat processed overseas. Hence, the last internationalization wave of the 1990s shows that the large Japanese food-processing firms are trying to keep pace with the general globalization process of the world's economy. Along with this somewhat peculiar internationalization trend, Japanese MNEs have developed new business units in their home market and have also changed from single-product firms to multiproduct/multidivision enterprises to increase control over their home market. This trend appears to be weakening their competitive international dynamics and threatens their future growth.

The composition of the pole termed "the rest of the world"[20] also changed during the last quarter of the twentieth century. Large industrial conglomerates left the world's agrofood landscape while others such as San Miguel Corporation (Philippines) and Goodman Fielder (Australia) did not attain the necessary increase in their size in order to stay in the world ranking. Among these MNEs, there has been a consolidation trend leading toward smaller size, achieved through refocusing on what they define to be their core businesses within the agrofood chain. However, for the moment, they do not seem to be sufficiently global to defend their presence within the world's food oligopoly. The recovery of Vivendi International (Canada) by the dismantling of Seagram is the last step in the necessary movement toward sectoral globalization.

Changes in the Composition of the Western European Pole of the Triad

British supremacy in the world's food oligopoly has diminished over time (Figure 1.7). British food MNEs, pioneers of international expansion in the nineteenth century, have decreased in number from twenty-one in 1974, to eighteen in 1988, to only ten in 2000. Their av-

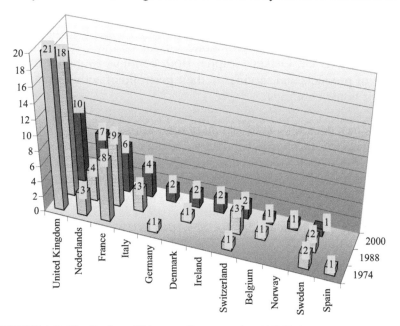

FIGURE 1.7. Distribution of Western European food MNEs by their countries of origin. (*Source:* Agrodata database, CIHEAM-IAMM, Montpellier.)

erage size, however, increased from US$1 billion in 1974 to US$5.4 billion in 2000. The fact that, in this study, Unilever is considered to be among the Dutch MNEs somewhat changes the composition of the Western European food oligopoly. Nevertheless, the present decrease in the number of British food MNEs has its roots in the economic crisis that Britain went through during the past two decades. The changing functions and structures of food-processing MNEs throughout the twentieth century is another factor explaining this decline. In fact, at the beginning of the twentieth century, multinational firms were located in foreign countries with comparative advantages regarding their agricultural raw materials. Because of the United Kingdom's political domination over Commonwealth countries, British food MNEs had some advantage over MNEs originating in other developed countries. Over time, the conquest of new consumer markets became the single most important factor leading to international expansion. The most dynamic and innovative MNEs won over those that placed their competitive advantage merely on raw material procurement. They are now being replaced by medium-sized MNEs, highly specialized in their operating sectors and originating in "small" European countries. Of course, this rule has some exceptions, such as the Norwegian industrial conglomerate Orkla, or the German multiproduct MNE Dr. Oetker Gruppe. Both continue to gather a wide range of products and services in their portfolio of activities, mainly in their home countries.

One of the events that modeled the European food oligopoly during the past decade was the arrival of medium-sized Dutch or Danish MNEs operating in new innovative sectors such as nutraceuticals and/or functional foods (Numico, CSM, Danisco) as well as in aquaculture and fish feed (Nutreco). Also of major importance is the extraordinary growth of Switzerland's MNE Nestlé, which increased its average size from US$5.6 billion in 1974 to US$24.6 billion in 2000. European brewing MNEs have also exhibited positive change dynamics since 2000. This is especially true with the Belgian brewer Interbrew, a small, country-based multinational at the end of the 1980s, which made significant growth by absorbing both Canadian John Labatt and the brewing business arms of Whitbread and Bass Plc. Meanwhile, it pursued an aggressive takeover policy toward ECE countries. It also got rid of its Belgian-based activities including bottled water, the soft drink business, and the Belgian franchising contract for bottling Coca-Cola. Already highly internationalized, Heineken also concentrated its company's resources around brewing, defined as its main business activity. It simultaneously applied an aggressive expansion strategy into the former socialist European countries. Another example of an MNE adopting expansion strategies is that of the British firm Scottish and Newcastle Breweries. However, being confined to the single European market, it seems to be suffering from market narrowness in spite of having bought Kronenbourg, the second largest brewer in the Western European Eureopan market, from Groupe Danone.

Changes in agricultural politics, such as the instauration of milk quotas in EU member countries since the beginning of the 1980s, have also acted as important macroeconomic factors influencing the reshaping of the European food oligopoly (Tozanli, 2001). Many dairy cooperatives and private enterprises facing these constraints merged and changed their market strategies as well as their juridicial status in order to attain the size crucial for international competitiveness.

CONCLUDING REMARKS

Using empirical evidence, I have shown the move toward globalization being made by many food-processing MNEs. It is no longer the company size or the breadth of activities in its portfolio that determines the efficiency of a multinational food-processing enterprise in the world market. It is rather a delicate combination of its core activities and their geographical spread over the world that matters. This change requires a centralized, vertical coordination of the MNEs' activities worldwide and pushes them to reshape their internal structures to become matrix-form organizations. In this manner, they redefine their relations with their subsidiaries within their business network. Hence, both the cultural and the decision-making pattern of the MNE become global as well.

These assessments remain partial. All these changes require a much more in-depth analysis. Governance structures, agent-principal relations, constraints imposed by stake and shareholders, and economic and financial performance must be integrated into a widened analysis so as to describe this new phenomenon of global enterprise in the world's food chain. This initial description of the globalization of food MNEs through their market strategies must be cross-referenced with their economic and financial performance in order to support or refute the link between geographical and sectoral globalization indicators (see Chapter 2).

This concentration process does not preclude the continuing existence of a multitude of small and medium-sized food processing firms (SMEs) which themselves are experiencing an ongoing concentration process. I am of the opinion that a certain amount of specialization is unavoidable in this globalization process. Consequently, large MNEs globalize in sectors needing high technology as well as standardized food products, while SMEs, especially in Western markets, respond more and more to a demand for local, quality food products. The future growth trend of these local SMEs depends upon their ability to take advantage of this new opportunity.

APPENDIX

The Agrodata database was developed at CIHEAM-IAM, Montpellier, in the 1970s. It provides information on the largest food-processing multinational firms worldwide. In addition to financial data based on the consolidated financial statements of multinational corporations, it contains information about these MNEs distributed by international press or published in their annual reports. This information has been quantified to draw up tables on restructuring operations (mergers, acquisitions, joint ventures, etc.) as well as on subsidiaries of these MNEs. Original analysis on strategic orientations and on financial and economic performance of these MNEs can now be realized thanks to the compilation of these original data. All the data used by the author to create the original tables and figures in this chapter has been extracted from this database.

TABLE A1.1. Distribution of acquisitions and majority shares according to the host zone of the operation and to the sector of business activities, January 1, 1987, to January 31, 2002.

Business activities	Africa	Latin America	Asia	ECE	Mediter-ranean	Japan	Oceania	North America	Western Europe	World
Upstream activities	**1**	**5**	**2**	**1**			**5**	**28**	**43**	**85**
Food processing										
HVA products		13	12	6	3	1	4	65	101	205
Milk, butter	6	19	6	20	4		4	34	102	197
Multiproducts	1	19	15	6	1	1	9	62	67	182
Fruit and vegetables	2	10	3	7	1		2	51	83	159
Meat processing		6	3	2			11	43	67	132
Chocolate and confectionery	1	8	5	13	2	1	1	20	44	95
Baking products		4			1		4	30	53	92
Edible fats and oils	1	8	4	9	4	1	1	13	38	79
Cheese		5		8			2	14	37	66
Biscuits		9	2	7	1		5	6	29	59
Convenience food		1	1	1	1			19	33	56
Frozen food		2		1			1	13	39	56
Animal feed	1	3	1	5				15	24	49
Microbiotics, proteins, nutraceuticals		4	2	1			1	11	25	44

29

TABLE A1.1 (continued)

Business activities	Africa	Latin America	Asia	ECE	Mediter-ranean	Japan	Oceania	North America	Western Europe	World
Snack foods			2	3	2		3	11	18	39
Ice cream		9	5	2	1			7	14	38
Fish processing				1				15	18	34
Sugar		2	1	12	1		1	3	10	30
Grain milling		3					4	11	11	29
Pet food		2					1	12	14	29
Yogurt, dairy desserts		1		5	2			4	13	26
Pasta products		2	1		1		1	4	14	23
Breakfast cereals			1					4	4	9
Total	12	130	64	109	25	5	55	467	858	1,728
Beverages										
Beer	5	7	13	46	2		9	12	51	145
Spirits	1	7	4	7			2	11	49	81
Soft drinks	2	11	4	4	4			25	27	77
Bottled water	3	2	3	10		1	2	8	9	39
Wine		4					4	8	17	33
Total	11	31	24	67	6	1	17	64	153	375
Tobacco		2		4	1	1		3	2	*11*
Downstream activities										
Food trading	2	8	1	1	1	1	2	49	73	139

									Total
Away-from-home eating, catering	2	4					1	22	32	61
Bottling		6	3	3	1		2	26	5	47
Food retailing		2						21	13	36
Packaging		2		1			1	4	7	15
Financing, insurance, real estate		1					1	3	6	11
R&D						2			4	6
Marketing		1			2				1	4
Transportation									4	4
Storage activities			1						2	3
Total	2	22	7	5	4	4	7	127	147	326
Nonfood activities	14	15	4	4	3	3	3	53	65	162
Diversified activities		1	1	1				3	1	6
Nonidentified	1	2	2					6	10	19
Total	41	205	104	190	36	13	87	751	1,279	2,712

TABLE A1.2. Distribution of mergers, joint ventures, license and franchising agreements, minority shares according to the host zone of the operation and to the sector of business activities, January 1, 1987, to January 31, 2002.

Business activity	Africa	Latin America	Asia	ECE	Mediter-ranean	Japan	Oceania	North America	Western Europe	World
Upstream activities							2	14	20	**36**
Food processing										
Multiproducts		2	1	1			2	27	31	64
Fruit and vegetables	2	1		1				15	41	60
Meat processing		1	1				4	5	46	57
Milk, butter	1		2	2		2	1	10	38	56
HVA products		3	1				2	19	30	55
Baking products		1	1					13	22	37
Chocolate and confectionery	1		1	2				7	23	34
Biscuits		4		1	1			5	15	26
Frozen food				1			1	4	16	23
Fish processing		2		1				4	14	21
Ice cream		1		1		1	1	1	15	20
Sugar	2	1		1			1	3	10	18
Animal feed		1				1		3	12	17
Cheese		1						5	11	17
Edible fats and oils								3	13	16

Grain milling	16	9	4			1			1	6
Snack foods	15	11	3				1			
Convenience food	14	10	4							
Microbiotics, proteins, nutraceuticals	13	7	4	1					1	
Pasta products	12	9	1				1		1	
Yogurt, dairy desserts	10	10								
Breakfast cereals	5	4					1			
Pet food	1	1								
Total	607	398	140	14	4	1	14	8	21	6
Beverages										
Beer	31	19	1	3	4		1	2		
Bottled water	8	4	3							
Soft drinks	8	4	4							
Spirits	45	35	6	1					1	
Wines	30	18	11	1						
Total	122	80	25	5	4			2	1	
Tobacco	10	3	3						2	1
Downstream activities										
Away-from-home eating, catering	65	44	20		1					
Food retailing	20	15	4					1		
Food trading	32	22	8	2						

TABLE A1.2 (continued)

Business activity	Africa	Latin America	Asia	ECE	Mediter-ranean	Japan	Oceania	North America	Western Europe	World
R&D								2	7	9
Bottling								2	11	13
Financing, insurance, real estate								5	6	11
Packaging								4	5	9
Storage activities									2	2
Transportation								1	2	3
Total			1			1	2	46	114	164
Nonfood activities							3	44	53	100
Diversified activities								4	8	12
Nonidentified	2							6	3	11
Total (numbers)	10	23	11	15	1	9	26	282	679	1,062

TABLE A1.3. Distribution of sell-offs, divestitures, and plant closures according to the host zone of the operation and to the sector of business activities, January 1, 1987, to January 31, 2002.

Business activity	Africa	Latin America	Asia	ECE	Mediter-ranean	Japan	Oceania	North America	Western Europe	World
Upstream activities							**2**	**14**	**20**	**36**
Food processing										
Multiproducts		2	1	1			2	27	31	64
Fruit and vegetables	2	1		1				15	41	60
Meat processing		1	1				4	5	46	57
Milk, butter	1		2	2		2	1	10	38	56
HVA products		3	1				2	19	30	55
Baking products		1	1					13	22	37
Chocolate and confectionery	1		1	2				7	23	34
Biscuits		4		1	1			5	15	26
Frozen food				1			1	4	16	23
Fish processing		2		1				4	14	21
Ice cream		1		1		1	1	1	15	20
Sugar	2	1		1			1	3	10	18
Animal feed		1				1		3	12	17
Cheese		1						5	11	17
Edible fats and oils								3	13	16

TABLE A1.3 *(continued)*

Business activity	Africa	Latin America	Asia	ECE	Mediter-ranean	Japan	Oceania	North America	Western Europe	World
Grain milling		1	1				1	4	9	16
Snack foods				1				3	11	15
Convenience food								4	10	14
Microbiotics, proteins, nutraceuticals		1					1	4	7	13
Pasta products		1		1				1	9	12
Yogurt, dairy desserts									10	10
Breakfast cereals				1					4	5
Pet food									1	1
Total	6	21	8	14	1	4	14	140	398	607
Beverages										
Spirits			2	1		4	3	1	19	31
Beer								3	4	8
Wines								4	4	8
Bottled water	1						1	6	35	45
Soft drinks			2				1	11	18	30
Total	1		2	1		4	5	25	80	122
Tobacco	1	2						3	3	10
Downstream activities										
Away-from-home eating, catering						1		20	44	65

Food retailing			1					4	15	20
Food trading							2	8	22	32
R&D								2	7	9
Bottling								2	11	13
Financing, insurance, real estate								5	6	11
Packaging								4	5	9
Storage activities									2	2
Transportation						1		1	2	3
Total			1		1		2	46	114	164
Nonfood activities							3	44	53	100
Diversified activities								4	8	12
Nonidentified	2							6	3	11
Total	10	23	11	15	1	9	26	282	679	1,062

TABLE A1.4. Distribution of restructuring operations realized by the top 100 according to main geographical zones, January 1, 1987, to January 31, 2002.

Home zone of the MNE	Africa	Latin America	North America	Japan	Asia	ECE	Mediter-ranean	Oceania	Western Europe	Home country of the parent company	Total op-erations
Number of acquisitions and majority takeovers (more than 50 percent of the capital of the target firm)											
Before 1992											
Western Europe	4	19	140	1	3	7	1	19	267	222	684
United States	1	18	28	1	1	1	–	4	84	200	339
Japan	–	–	10	–	2	–	–	5	7	2	26
Rest of the world	–	2	13	–	2	–	–	5	21	22	66
Top 100	5	39	191	2	8	8	1	33	379	446	1,115
After 1992											
Western Europe	17	100	133	4	56	136	20	19	306	243	1,051
United States	3	55	17	1	25	34	2	18	126	179	471
Japan	–	–	9	–	2	–	1	3	6	6	25
Rest of the world	2	9	16	–	11	10	–	2	11	24	86
Top 100	22	164	175	5	94	180	23	42	459	452	1,633

Number of mergers, minority takeovers (less than 50 percent of the capital of the target firm), partnerships, licences, and franchising contracts

Before 1992

Western Europe	7	9	27	13	26	15	4	6	124	95	327
United States	–	4	2	4	16	18	4	2	28	28	106
Japan	–	1	9	–	14	1	–	5	19	18	67
Rest of the world	–	1	5	1	1	–	–	3	14	12	37
Top 100	7	15	43	18	57	34	8	16	185	153	537

After 1992

Western Europe	4	26	30	9	69	64	22	10	105	79	420
United States	5	19	1	3	11	14	8	3	38	45	150
Japan	–	–	3	–	22	3	–	–	8	7	43
Rest of the world	–	3	3	–	4	3	–	1	4	6	25
Top 100	9	48	37	12	84	84	30	14	155	137	638

TABLE A1.4 (continued)

Home zone of the MNE	Africa	Latin America	North America	Japan	Asia	ECE	Mediter-ranean	Oceania	Western Europe	Home country of the parent company	Total operations
Number of sell-offs, spin-offs, divestitures											
Before 1992											
Western Europe	1	2	68	2	3	–	–	5	66	193	340
United States	2	4	9	–	1	–	–	3	45	95	159
Japan	–	–	–	–	–	–	–	–	–	–	–
Rest of the world	1	–	16	2	4	–	–	3	6	22	51
Top 100	4	6	93	4	8	–	–	11	117	310	550
After 1992											
Western Europe	7	11	54	4	8	9	1	8	188	198	488
United States	2	18	5	1	6	10	1	6	49	111	209
Japan	–	–	1	–	–	–	–	2	0		3
Rest of the world	–	–	6	5	14	–	–	2	16	5	48
Top 100	9	29	66	10	28	19	2	18	253	314	748

Number of plant closures (after 1990)

	1	2	3	4	5	6	7	8	Total
Netherlands	1	3	–	–	–	–	18	8	30
Norway	–	–	–	–	–	–	–	1	1
Switzerland	–	–	2	–	–	–	7	1	10
Germany	–	–	1	–	–	–	1	–	2
Belgium	–	–	–	–	–	–	1	–	1
France	–	–	–	–	1	–	5	10	16
United Kingdom	–	–	–	–	–	–	3	23	26
Ireland	–	–	–	–	–	–	–	1	1
Italy	–	–	–	–	–	–	3	–	3
Western Europe	–	3	3	–	1	–	38	44	90
United States	–	4	1	–	–	–	9	10	24
Japan	–	–	–	–	–	–	–	6	6
Canada	–	–	–	–	–	–	1	1	2
Top 100	–	7	4	–	1	–	47	60	120

TABLE A1.5. World ranking of the leading food-processing multinational enterprises (in current U.S. dollars in 2000).

Rank	MNE name	Home country	Sector of main business activity	Total sales millions US$	Agrifood sales millions US$	Net income millions US$	Salaried workers
1	Nestlé	Switzerland	Multiproducts	48,210	45,369	3,412	224,541
2	Philip Morris (Altria)	United States	Multiproducts	63,276	29,977e	8,510	178,000
3	Unilever	Netherlands	Multiproducts	44,254	22,084	1,215	295,000
4	ConAgra	United States	Multiproducts	25,386	21,900	382	89,000
5	Cargill Inc.	United States	Grain milling	48,000	21,000	500	90,000
6	Coca-Cola Co.	United States	Soft drinks	20,458	20,458	2,177	36,900
7	PepsiCo Inc.	United States	Soft drinks, snack foods	20,438	20,438	2,183	53,000
8	IBP Inc.	United States	Meat processing	16,950	16,950	153	50,000
9	Mars Inc.	United States	Confectionery, chocolates	15,300	15,300e	1,350	30,000
10	Diageo Plc	United Kingdom	Wine and spirits	15,179	15,179	1,918	66,668
11	Archer Daniels Midland	United States	Edible fats and oils	18,612	13,972	301	22,834
12	Groupe Danone	France	Multiproducts	13,201	13,201	666	86,657
13	Snow Brand Milk Products	Japan	Dairy products	11,976	10,515e	265	15,127
14	Suntory	Japan	Wine and spirits	12,018	10,114e	nc	4,873
15	Anheuser Busch Inc.	United States	Beer	12,262	9,560	1,552	23,725
16	H.J. Heinz Co.	United States	Multiproducts	9,408	9,408	891	45,800
17	Montedison (EBS)	Italy	Edible fats and oils	13,216	9,060e	240	nc

18	Bunge & Born	Argentina	Grain milling	11,000	9,000e	nc	nc
19	Maruha Corp.	Japan	Fishing, fish processing	8,754	8,195e	21	1,064
20	Nippon Meat Packers	Japan	Meat processing	8,175	8,175	141	nc
21	Kirin Brewery Co.	Japan	Beer	8,862	7,993	309	21,867
22	Asahi Breweries	Japan	Beer	7,722	7,707	–1	4,103
23	Tyson Foods	United States	Poultry	7,410	7,410	151	124,000
24	Kellogg Company	United States	Breakfast cereals	6,955	6,955	588	15,200
25	Cadbury Schweppes	United Kingdom	Soft drinks, confectionery	6,936	6,936	1,237	36,460
26	Yamazaki Baking	Japan	Baking products, biscuits	6,822	6,822	76	nc
27	Parmalat Finanziaria SpA	Italy	Dairy products	6,790	6,790	180	38,303
28	General Mills Inc.	United States	Multiproducts	6,700	6,700	614	11,077
29	Associated British Foods Plc	United Kingdom	Multiproducts	6,680	6,680	206	34,372
30	Heineken NV	Netherlands	Beer	6,481	6,481	574	37,857
31	Meiji Dairies	Japan	Dairy products	6,408	6,408	43	8,681
32	Campbell Soup	United States	Multiproducts	6,466	6,376	714	24,000
33	Farmland Industries	United States	Grain milling	12,239	5,943	135	14,500
34	Ajinomoto	Japan	Multiproducts	7,714	5,785	159	22,373
35	Suiza Foods	United States	Dairy products	5,756	5,756	144	18,277
36	Interbrew	Belgium	Beer	5,227	5,227	300	34,203
37	Smithfield Foods	United States	Meat processing	5,150	5,150	81	30,000
38	The Seagram Co. Inc.	Canada	Wine and spirits	15,686	5,108	40	34,000

TABLE A1.5 (continued)

Rank	MNE name	Home country	Sector of main business activity	Total sales millions US$	Agrifood sales millions US$	Net income millions US$	Salaried workers
39	Tate & Lyle Plc	United Kingdom	Sugar, sweeteners	5,428	5,100	403	20,085
40	Sara Lee Corporation	United States	Multiproducts	17,511	5,088	2,041	154,200
41	Quaker Oats Co.	United States	Multiproducts	5,041	5,041	361	11,858
42	Lactalis (Besnier S.A.)	France	Dairy products	4,943	4,943	0	16,000
43	Morinaga Milk Industry	Japan	Dairy products	4,776	4,776	-24	nc
44	Sapparo Breweries	Japan	Beer	4,942	4,685	4,097	5,317
45	Procter & Gamble	United States	Multiproducts	39,951	4,634	3,542	106,000
46	Uniq (Unigate)	United Kingdom	Dairy products	4,514	4,514	141	37,813
47	Arla Foods	Denmark	Dairy products	4,755	4,481	144	18,600
48	Nichirei Corp.	Japan	Fishing, fish processing	5,296	4,476	40	6,907
49	Dole Foods Co. Inc.	United States	Fruits and vegetables processing	4,763	4,418	68	61,000
50	Ito Ham Foods Inc.	Japan	Meat processing	4,375	4,375	59	nc
51	Land O'Lakes Inc.	United States	Dairy products	5,756	4,364	103	6,500
52	Hershey Foods Inc.	United States	Confectionery, chocolates	4,221	4,221	364	14,300
53	Sudzucker	Germany	Sugar, sweeteners	4,173	4,173	160	29,579
54	Dean Foods	United States	Dairy products	4,103	4,103	106	14,050
55	Mc Cain Foods	Canada	Fruits and vegetables processing	4,100	4,100	nc	16,000
56	Pernod Ricard	France	Wine and spirits	4,049	4,049	185	14,609
57	Friesland Dairies	Netherlands	Dairy products	4,068	4,028	78	11,970

58	Ferrero SpA	Italy	Confectionery, chocolates	4,000	4,000e	nc	14,000
59	Nippon Suisan Kaisha	Japan	Fishing, fish processing	4,392	3,950	47	nc
60	Numico	Netherlands	Baby food, nutraceutics	3,903	3,903	347	31,523
61	Novartis	Switzerland	Functional foods, baby food	21,200	3,786	646	67,653
62	Hormel Foods	United States	Meat processing	3,675	3,675	170	12,200
63	Scottish & New-castle Breweries	United Kingdom	Beer	4,270	3,655	275	57,745
64	Wessanen	Netherlands	Dairy products	3,635	3,635e	69	17,706
65	Campina Melkuni	Netherlands	Dairy products	3,598	3,598	22	7,615
66	Bongrain	France	Dairy products	3,580	3,580	53	14,751
67	Q.P. Corporation	Japan	Highly processed food	3,574	3,574	-12	6,220
68	Nisshin Flour Milling	Japan	Grain milling	3,747	3,433	101	2,450
69	South African Breweries	South Africa	Beer	4,715	3,327	578	48,079
70	Carlsberg A/S	Denmark	Beer	3,206	3,206	202	23,691
71	Kikkoman	Japan	Highly processed food	3,038	3,038	57	6,555
72	Nutreco	Netherlands	Animal feed, aquaculture	2,888	2,888	83	9,603
73	Toyo Suisan Kaisha	Japan	Highly processed food	2,881	2,881	44	nc
74	Tomkins Plc	United Kingdom	Grain milling, baking	8,517	2,879	512	70,019
75	Contigroup	United States	Poultry	4,340	2,800e	nc	14,500
76	Ralston Purina	United States	Baking products, biscuits	2,763	2,763	530	4,380
77	Nission Food Products	Japan	Highly processed food	2,721	2,721	144	5,368

TABLE A1.5 *(continued)*

Rank	MNE name	Home country	Sector of main business activity	Total sales millions US$	Agrifood sales millions US$	Net income millions US$	Salaried workers
78	Whitbread & Co. Plc	United Kingdom	Beer	4,475	2,688	273	73,058
79	Ambev	Brazil	Beer	2,686	2,685	224	17,500
80	Allied Domecq Plc	United Kingdom	Wine and spirits	3,119	2,656	494	10,932
81	Maple Leaf Foods Inc.	Canada	Animal feed	2,628	2,628	25	14,500
82	Sodiaal	France	Dairy products	2,549	2,549	nc	nc
83	Oetker Gruppe (+ Binding)	Germany	Multiproducts	4,793	2,500	nc	17,763
84	Meiji Seika Kaisha	Japan	Confectionery	3,379	2,440e	42	7,430
85	Ezaki Glico	Japan	Confectionery	2,426	2,426	24	4,451
86	Danisco A/S	Denmark	Sugar, sweeteners, nutraceutics	3,470	2,401e	109	17,712
87	Kerry Group Plc	Ireland	Food ingredients	2,423	2,386	122	13,410
88	International Multifoods	United States	Multiproducts	2,385	2,385	25	4,654
89	CSM	Netherlands	Food ingredients, nutraceutics	2,518	2,342	139	10,905
90	Constellation Brands (Canandaguia)	United States	Wine and spirits	2,341	2,341	77	4,500
91	Orkla	Netherlands	Multiproducts	3,943	2,309e	431	31,145
92	Nichiro Gyogyo Kaisha	Japan	Fishing, fish processing	2,252	2,252	30	nc
93	Chiquita Brands International	United States	Fruits and vegetables processing	2,253	2,245	–95	30,000
94	Glanbia	Ireland	Dairy products	2,219	2,219	24	7,367

46

95	Barilla SpA	Italy	Baking and pasta products, biscuits	2,173	2,173	93	8,500
96	LVMH	France	Wine and spirits	10,701	2,158	667	51,127
97	Wm. Wrigley Jr. Company	United States	Confectionery	2,146	2,146	329	9,800
98	MC Cormick and Co	United States	Highly processed food	2,124	2,124	138	8,100
99	Earthgrains	United States	Baking	2,039	2,039	55	26,000
100	Northern Foods	United Kingdom	Dairy products	2 030	2,030	86	2,200
101	Corn Foods International	United States	Grain milling	1,865	1,865	48	6,000
102	Goodman Fielder	Australia	Grain milling, baking	1,826	1,826	88	14,248
103	Gold Kist	United States	Poultry	1,770	1,743	-26	18,000
104	Moksel AG	Germany	Meat processing	1,740	1,740	2	2,421
105	Flower Foods	United States	Baking	1,619	1,619	-42	7,300
106	Brown Forman	United States	Wine and spirits	2,134	1,543	218	7,400
107	Fromageries Bel	France	Dairy products	1,514	1,514	64	8,000
108	San Miguel Corporation	Philippines	Beer, soft drinks	2,005	1,470	170	14,511

e: estimated; nc: not communicated

TABLE A1.6. World ranking of the 100-top food-processing multinational enterprises (in current US dollars in 1988).

Rank	MNE name	Home country of the parent company	Sector of main business activity	Total sales (millions US$)	Agrofood sales (millions US$)	Net income (millions US$)	Salaried workers
1	Cargill Inc.	United States	Grain milling	38,800	38,800	nc	47,000
2	Nestlé	Switzerland	Confectionery, chocolates	27,800	26,600	1,390.3	197,700
3	Unilever	Netherlands	Edible fats and oils	31,303	18,835	1,526.2	291,000
4	Philip Morris	United States	Multiproducts	25,860	14,527	2,337.0	155,000
5	PepsiCo Inc.	United States	Soft drinks	13,007	13,007	762.2	235,000
6	RJR Nabisco	United States	Multiproducts	16,956	9,835	1,393.0	116,881
7	Grand Metropolitan	United Kingdom	Wine and spirits	10,726	9,793	733.0	89,753
8	Elders IXL	Australia	Agribusiness	11,209	9,128	578.4	38,000
9	Bunge & Born	Argentina	Grain milling	18,000	9,000	nc	nc
10	Con Agra	United States	Grain milling	9,475	8,800	154.7	42,993
11	Mars Inc.	United States	Confectionery, chocolates	9,200	8,785	nc	nc
12	George Weston	Canada	Food trading	8,869	8,618	156.0	61,400
13	Anheuser Busch Co.	United States	Beer	8,924	8,583	715.9	41,118
14	Coca-Cola Co.	United States	Soft drinks	8,338	8,300	1,044.7	22,000
15	Allied Lyons	United Kingdom	Beer	8,006	8,006	558.1	53,166

16	Dalgety Inc.	United Kingdom	Multiproducts	7,969	7,787	142.3	22,820
17	Sara Lee Corp.	United States	Multiproducts	10,424	7,351	325.1	85,700
18	Taiyo Fishery	Japan	Fishing, fish processing	8,850	7,330	65.9	nc
19	Gruppo Ferruzzi	Italy	Sugar, sweeteners	19,079	6,841	424.8	85,210
20	Snow Brand Milk Products	Japan	Dairy products	7,139	6,425	52.2	nc
21	BSN Groupe	France	Multiproducts	7,065	6,383	366.7	42,234
22	Suntory	Japan	Wine and spirits	6,371	6,371	62.2	
23	Pillsbury	United States	Multiproducts	6,191	6,191	69.3	nc
24	Archer Daniels Midland	United States	Edible fats and oils	6,798	6,100	353.1	9,631
25	Borden Inc.	United States	Dairy products	7,244	5,381	311.9	46,300
26	H. J. Heinz	United States	Multiproducts	5,244	5,244	386.0	39,000
27	General Mills	United States	Breakfast cereals	5,179	5,179	283.0	74,500
28	Kirin Brewery Co. Ltd.	Japan	Beer	5,100	5,100	253.9	7,582
29	Campbell Soup	United States	Fruits and vegetables processing	4,869	4,869	274.1	48,389
30	Groupe Sucre et Denrée	France	Diversified	5,495	4,858	11.7	5,800
31	Ralston Purina Co.	United States	Biscuits, baking and pasta products	5,876	4,737	387.8	56,734
32	CPC International	United States	Grain milling	4,700	4,700	289.1	32,000

TABLE A1.6 *(continued)*

Rank	MNE name	Home country of the parent company	Sector of main business activity	Total sales (millions US$)	Agrofood sales (millions US$)	Net income (millions US$)	Salaried workers
33	Guinness PLC	United Kingdom	Wine and spirits	4,939	4,700	606.7	19,759
34	Nippon Meat Packers	Japan	Meat processing	4,604	4,604	89.5	nc
35	Quaker Oats	United States	Breakfast cereals	5,330	4,508	255.7	31,300
36	Beatrice Co. Inc.	United States	Multiproducts	4,498	4,498	325.0	16,400
37	Hillsdown Holdings	United Kingdom	Meat processing	6,320	4,497	368.1	42,832
38	Bass PLC	United Kingdom	Beer	6,644	4,446	574.3	84,843
39	Associated British Foods	United Kingdom	Biscuits, baking and pasta products	4,398	4,398	345.6	53,024
40	Jacobs Suchard	Switzerland	Confectionery, chocolates	4,354	4,354	234.3	16,799
41	Kellogg Co.	United States	Breakfast cereals	4,349	4,349	480.4	17,461
42	Cadbury Schweppes	United Kingdom	Confectionery, chocolates	4,241	4,241	300.8	28,874
43	United Biscuits	United Kingdom	Biscuits, baking and pasta products	4,239	4,239	291.2	47,004
44	John Labatt	Canada	Beer	3,959	3,959	109.0	17,900
45	Nippon Suisan Kaisha Ltd	Japan	Fishing, fish processing	4,138	3,807	21.5	nc
46	Sapporo Breweries	Japan	Beer	3,945	3,748	50.0	3,884
47	Seagram Co. Ltd.	Canada	Wine and spirits	3,710	3,710	589.5	16,200

	Company	Country	Business				
48	Heineken N.V.	Netherlands	Beer	3,686	3,683	146.9	28,719
49	Tate & Lyle PLC	United Kingdom	Sugar, sweeteners	3,715	3,641	124.9	14,040
50	S & W Berisford Ltd	United Kingdom	Multiproducts	4,706	3,558	59.1	9,108
51	BP Nutrition	United Kingdom	Animal feed	3,514	3,514	nc	11,850
52	Barlow Rand	South Africa	Diversified	9,774	3,415	611.0	157,500
53	United Brands Co.	United States	Fruits and vegetables processing	3,503	3,400	60.4	42,000
54	Whitbread & Co.	United Kingdom	Beer	3,280	3,280	286.9	26,954
55	Ito Ham Foods	Japan	Meat processing	3,161	3,161	54.3	4,528
56	Unigate PLC	United Kingdom	Dairy products	4,149	3,113	229.7	35,195
57	S.M.E.	Italy	Multiproducts	3,099	3,099	68.1	nc
58	Ajinomoto	Japan	Multiproducts	3,962	3,086	122.7	9,532
59	Yamazaki Baking	Japan	Biscuits, baking and pasta products	3,202	2,983	70.9	15,919
60	Rank Hovis Mc Dougall	United Kingdom	Grain milling	2,969	2,956	186.3	38,195
61	Booker PLC	United Kingdom	Agribusiness	3,275	2,938	130.5	17,166
62	Nichirei Corporation	Japan	Fishing, fish processing	3,264	2,913	33.0	nc
63	Procter & Gamble	United States	Diversified	19,336	2,863	1,020.0	77,000
64	Nisshin Flour Milling	Japan	Grain milling	2,799	2,758	60.0	nc
65	Bond Corp. Holdings Ltd.	Australia	Diversified	3,657	2,556	199.7	10,408
66	Source Perrier	France	Soft drinks	2,443	2,443	175.4	17,804

TABLE A1.6 *(continued)*

Rank	MNE name	Home country of the parent company	Sector of main business activity	Total sales (millions US$)	Agrofood sales (millions US$)	Net income (millions US$)	Salaried workers
67	Whitman (IC Industries)	United States	Diversified	3,583	2,352	233.5	25,396
68	Geo Hormel & Co.	United States	Meat processing	2,293	2,293	60.2	7,994
69	Canada Packers Inc.	Canada	Meat processing	2,664	2,279	20.8	12,000
70	Castle & Cooke	United States	Fruits and vegetables processing	2,469	2,268	112.3	42,000
71	Pernod Ricard	France	Wine and spirits	2,256	2,256	130.0	10,287
72	Hershey Foods Corp.	United States	Confectionery, chocolates	2,168	2,168	213.9	12,100
73	Sodima	France	Dairy products	2,144	2,144	nc	7,200
74	Nichiro Gyogyo Kaisha	Japan	Fishing, fish processing	2,174	1,981	10.6	2,855
75	Tyson Foods	United States	Meat processing	1,936	1,936	81.4	26,000
76	Wessanen N.V.	Netherlands	Dairy products	1,923	1,923	46.8	6,571
77	Morinaga Milk Industry	Japan	Dairy products	2,592	1,919	29.7	4,304
78	International Multifoods	United States	Multiproducts	1,874	1,874	35.1	9,015
79	Northern Foods	United Kingdom	Dairy products	1,834	1,834	90.7	21,864
80	Union Laitière Normande	France	Dairy products	1,819	1,819	14.4	5,690
81	Union International PLC	United Kingdom	Meat processing	1,819	1,819	31.6	23,161

	Company	Country	Sector				
82	Ferrero SpA	Italy	Confectionery, chocolates	1,765	1,765	nc	nc
83	Land O'Lakes	United States	Dairy products	2,252	1,730	30.4	6,724
84	Provendor Group	Sweden	Multiproducts	1,707	1,707	nc	9,300
85	Goodman Fielder Wattie	Australia	Multiproducts	1,726	1,700	nc	20,000
86	Ezaki Glico	Japan	Confectionery, chocolates	1,695	1,695	38.4	nc
87	Coberco Melk Industrie	Netherlands	Dairy products	1,688	1,688	965.2	3,337
88	MD Foods	Denmark	Dairy products	1,699	1,686	92.8	5,517
89	QP Corporation	Japan	Highly processed food	1,672	1,672	28.8	3,942
90	Sandoz	Switzerland	Multiproducts	6,925	1,648	519.2	48,079
91	Besnier	France	Dairy products	1,629	1,629	63.3	5,745
92	Arla	Sweden	Dairy products	1,591	1,579	nc	6,176
93	Dean Foods	United States	Dairy products	1,552	1,552	42.8	7,100
94	Kyokyo	Japan	Fishing, fish processing	1,658	1,492	4.1	nc
95	LVMH	France	Wine and spirits	2,754	1,484	335.5	12,355
96	Mc Cain Foods	Canada	Fruits and vegetables processing	1,474	1,474	nc	11,000
97	Kikkoman	Japan	Highly processed food	1,462	1,462	35.5	4,361
98	Artois Piedboeuf Interbrew	Belgium	Beer	1,439	1,439	nc	12,180
99	Toyo Suisan Kaisha	Japan	Fishing, fish processing	1,422	1,422	28.2	nc
100	Saint Louis	France	Sugar, sweeteners	1,419	1,412	84.8	7,988

TABLE A1.7. World ranking of the 100-top food-processing multinational enterprises (in current U.S. dollars in 1974).

Rank	MNEs	Home country of the parent company	Sector of main business activity	Total sales (millions US$)
1	Unilever	Netherlands	Edible fats and oils	13,667
2	Nestle Alimentana	Switzerland	Dairy	5,603
3	Swift (Esmark)	United States	Meat	4,616
4	Kraftco	United States	Dairy	4,471
5	Beatrice Foods	United States	Dairy	3,541
6	Greyhound (Armour)	United States	Meat	3,458
7	Borden	United States	Dairy	3,265
8	Ralston Purina	United States	Animal feed	3,073
9	General Foods	United States	Multiproducts	2,987
10	Cpc International	United States	Grains milling	2,570
11	Associated British Foods	United Kingdom	Bread and bakery products	2,526
12	Coca Cola	United States	Nonalcoholic beverages	2,522
13	Taiyo Fisheries	Japan	Fishing	2,490
14	Consolidated Foods	United States	Multiproducts	2,380
15	United Brands Amk	United States	Meat	2,230
16	Pepsico	United States	Soft drinks	2,081
17	Gervais Danone	France	Dairy	2,035
18	General Mills	United States	Breakfast cereals	2,000
19	Carnation	United States	Dairy	1,887
20	Nabisco	United States	Bread and bakery products	1,793

21	Central Soya	United States	Animal feed	1,749
22	CSR Colonial Sugar Refining	Australia	Sugar and sugar products	1,680
23	Ranks Hovis Mc Dougall	United Kingdom	Grains milling, bakery	1,652
24	Standard Brands	United States	Multiproducts	1,648
25	Norton Simon	United States	Multiproducts	1,600
26	Tate Lyle	United Kingdom	Sugar and sugar products	1,552
27	Archer Daniels Midland	United States	Edible fats and oils	1,551
28	Iowa Beef Processors	United States	Meat	1,537
29	Canada Packers	Canada	Meat	1,479
30	Campbell Soup	United States	Multiproducts	1,468
31	H. J. Heinz	United States	Multiproducts	1,438
32	Oetker Gruppe	Germany	Multiproducts	1,424
33	Associated Milk Producers	United States	Dairy	1,416
34	Anheuser Busch	United States	Beer	1,413
35	Union International	United Kingdom	Animal feed	1,405
36	Cavenham Ltd.	United Kingdom	Multiproducts	1,390
37	J. Lyons	United Kingdom	Bread and bakery products	1,366
38	Cadbury Schweppes	United Kingdom	Nonalcoholic beverages	1,299
39	Quaker Oats	United States	Breakfast cereals	1,227
40	Unigate	United Kingdom	Dairy	1,145
41	National Distillers	United States	Wine and spirits	1,089
42	Spillers	United Kingdom	Grains milling, protein concentrates	1,058
43	Amstar Corporation	United States	Sugar and sugar products	1,047

TABLE A1.7 *(continued)*

Rank	MNEs	Home country of the parent company	Sector of main business activity	Total sales (millions US$)
44	Groupe Cooperatif Gama	France	Dairy	1,046
45	Del Monte	United States	Fruits and vegetables	1,043
46	Beecham	United Kingdom	Nonalcoholic beverages	1,035
47	Svenska M. Riskforening	Sweden	Multiproducts	1,029
48	Snow Brand Milk Products	Japan	Dairy	1,012
49	Kellogg	United States	Breakfast cereals	1,010
50	Pillsbury	United States	Grains milling	1,004
51	Oscar Mayer	United States	Meat	972
52	Groupe Cooperatif Mac Mahon	United States	Grains milling	970
53	Heublein	United States	Wine and spirits	968
54	Geo Hormel	United States	Meat	943
55	Kirin Brewery	Japan	Beer	935
56	Brooke Bond Liebig	United Kingdom	Multiproducts	905
57	American Beef Packers	United States	Meat	897
58	Distillers	United Kingdom	Wine and spirits	889
59	Seagram	Canada	Wine and spirits	886
60	Ajinomoto	Japan	Multiproducts	885
61	Pet Inc.	United States	Edible fats and oils	883
62	Anderson Clayton	United States	Edible fats and oils	879
63	Bass Charington	United Kingdom	Beer	852
64	Died Breweries	United Kingdom	Beer	840

65	Jos Schiltz	United States	Beer	815
66	Reckitt and Colman	United Kingdom	Multiproducts	782
67	United Biscuits	United Kingdom	Grains milling	766
68	Castle and Cooke	United States	Fruits and vegetables	753
69	International Multifoods	United States	Bakery and pasta products	752
70	Missouri Beef Packers (Mbpxl)	United States	Meat	727
71	Molson Industry	Canada	Beer	708
72	Compagnie Financiere Lesieur	France	Edible fats and oils	691
73	Beghin-Say	France	Sugar and sugar products	688
74	Nisshin Flour	Japan	Grains milling	678
75	Booker Mc Connell	United Kingdom	Agribusiness	671
76	Meiji Milk Products	Japan	Dairy	662
77	Kane Miller	United States	Meat	661
78	Union Laitiere Normande	France	Dairy	651
79	Arthur Guinness	United Kingdom	Beer	641
80	Liggett and Myers	United States	Multiproducts	627
81	Koninklijke Wessanen	Netherlands	Animal feed	621
82	Sodima Yoplait	France	Dairy	621
83	Rumasa	Spain	Multiproducts	607
84	Rowntree Mackintosh	United Kingdom	Chocolates, confectionery	592
85	Campbell Taggart	United States	Bakery, biscuits	590
86	Burns Foods	Canada	Meat	582

TABLE A1.7 *(continued)*

Rank	MNEs	Home country of the parent company	Sector of main business activity	Total sales (millions US$)
87	Mjolkcentralen	Sweden	Dairy	550
88	Morinaga	Japan	Dairy	524
89	Pernod Ricard	France	Wine and spirits	515
90	Hershey Foods	United States	Chocolates, Confectionery	514
91	Di Giorgio Corp.	United States	Distribution of foodstuffs	508
92	Heineken	Netherlands	Beer	491
93	Scottish & Newcastle Breweries	United Kingdom	Beer	484
94	Whitbread	United Kingdom	Beer	484
95	Mars	United States	Chocolates, Confectionery	476
96	Ward Foods	United States	Bread and bakery products	473
97	Hygrade Foods	United States	Meat	472
98	Perrier	France	Soft drinks	466
99	Libby Mc Neil & Libby	United States	Fruits and vegetables	465
100	Hiram Walker-Goderham	Canada	Wine and spirits	455

TABLE A1.8. Refocusing on core business activities and sectoral globalization indicator (SGI)[a] of the leading food processing MNEs according to the geographical distribution of their subsidiaries in 1988 and 1999-2000.

Food MNE	Home zone of the parent company	Number of total business activities in overseas markets		Number of total business activities in home market		SGI (%)	
		1988	2000	1988	2000	1988	2000
Danone	Western Europe	2	3	9	3	22	100
Bongrain	Western Europe	1	1	3	1	33	100
Barilla	Western Europe	1	1	2	1	50	100
LVMH	Western Europe	1	1	2	1	50	100
Cadbury Schweppes	Western Europe	2	2	3	2	67	100
Pernod Ricard	Western Europe	3	2	3	2	100	67
Ferrero	Western Europe	1	1	1	1	100	100
Lactalis	Western Europe	1	1	1	1	100	100
Sodiaal	Western Europe	1	1	1	1	100	100
Diageo PLC	Western Europe	–	2	–	2	–	100
Ralston Purina	United States	1	2	4	2	25	100
Chiquita Brands	United States	1	1	4	1	25	100
Dole Food	United States	1	1	3	1	33	100
Coca-Cola	United States	1	1	2	1	50	100
International Multifoods	United States	2	2	2	2	100	100
Smithfield Foods	United States	–	1	–	1	–	100
Constellation Brands	United States	–	1	–	1	–	100

TABLE A1.8 (continued)

Food MNE	Home zone of the parent company	Number of total business activities in overseas markets		Number of total business activities in home market		SGI (%)	
		1988	2000	1988	2000	1988	2000
Earthgrains Company	United States	–	1	–	1	–	100
IBP Inc	United States	–	1	–	1	–	100
Nippon Suisan	Japan	2	2	2	2	100	100
Meiji Seika Kaisha	Japan	1	3	1	3	100	100
Seagram Co.	Rest of the world	1	1	3	1	33	100
Ambev	Rest of the world	–	1	–	1	–	100
Campbell Soup	United States	2	3	7	4	29	75
PepsiCo	United States	1	3	2	4	100	75
Novartis	Western Europe	2	2	3	3	67	67
Kellogg Company	United States	1	2	4	3	25	67
Mars Inc.	United States	2	2	4	3	50	67
Morinaga Milk Industry	Japan	1	2	1	3	100	67
Unilever	Western Europe	4	3	13	5	31	60
Montedison (EBS)	Western Europe	3	3	7	5	43	60
H. J. Heinz	United States	3	3	9	5	33	60
Archer Daniels Midland	United States	3	3	7	5	43	60
General Mills	United States	2	3	3	5	67	60
Sara Lee Corporation	United States	3	4	7	7	43	57
Allied Domecq PLC	Western Europe	2	1	8	2	25	50
Heineken	Western Europe	1	1	3	2	33	50
Interbrew	Western Europe	1	1	3	2	33	50

Campina Melkunie	Western Europe	1	2	3	4	33	50
Koninklijke Wessanen	Western Europe	4	2	6	4	67	50
Carlsberg	Western Europe	1	1	2	2	50	50
Danisco	Western Europe	–	3	–	6	–	50
Friesland Dairy Foods	Western Europe	–	1	–	2	–	50
Numico	Western Europe	–	2	–	4	–	50
Parmalat	Western Europe	–	1	–	2	–	50
Anheuser Busch Inc.	United States	1	1	6	2	7	50
Philip Morris	United States	4	3	8	6	50	50
McCormick & Co.	United States	1	1	2	2	50	50
Quaker Oats Company	United States	1	2	2	4	50	50
Suiza Dairy Group	United States	–	1	–	2	–	50
Wm. Wrigley Jr. Co.	United States	–	1	–	2	–	50
Kikkoman	Japan	1	1	2	2	50	50
Nippon Meat Packers	Japan	1	1	2	2	50	50
Nichirei Corp	Japan	2	1	3	2	67	50
Nichiro Gyogyo	Japan	1	1	1	2	100	50
Nisshin Flour Milling	Japan	1	2	1	4	100	50
Toyo Suisan	Japan	1	2	1	4	100	50
Nestle	Western Europe	3	4	12	9	25	44
Tate & Lyle	Western Europe	2	2	6	5	33	40
Nutreco	Western Europe	–	2	–	5	–	40
Mc Cain Foods LTD	Rest of the world	3	2	6	5	50	40

TABLE A1.8 (continued)

Food MNE	Home zone of the parent company	Number of total business activities in overseas markets		Number of total business activities in home market		SGI (%)	
		1988	2000	1988	2000	1988	2000
Bunge International	Rest of the world	3	3	7	8	43	38
Arla Foods	Western Europe	ns	2	3	6	ns	33
Hormel Foods	United States	1	1	4	3	25	33
Land o'Lakes	United States	1	1	3	3	33	33
Hershey Foods	United States	1	1	2	3	50	33
Farmland Industries	United States	–	1	–	3	–	33
Procter & Gamble	United States	–	1	5	3	20	33
Suntory	Japan	2	2	2	6	100	33
Ezaki-Glico	Japan	1	1	3	3	33	33
Nissin Food Products	Japan	1	1	3	3	33	33
Ito Ham Foods	Japan	1	1	2	3	50	33
Sapporo Breweries	Japan	1	1	2	3	50	33
Yamazaki Baking	Japan	–	1	1	3	–	33
Asahi Breweries	Japan	–	1	–	3	–	33
South African Breweries	Rest of the world	–	1	–	3	–	33
Associated British Foods	Western Europe	3	2	7	7	43	29
Kirin Brewery Co. LTD.	Japan	2	2	4	7	50	29
Glanbia	Western Europe	–	1	–	4	–	25
CSM	Western Europe	–	1	–	4	–	25
Oetker Gruppe	Western Europe	–	1	–	4	–	25

Company	Region						
Tyson Foods Inc.	United States	1	1	4	4	25	25
Contigroup	United States	–	1	–	4	–	25
Dean Foods	United States	1	1	3	4	33	25
QP Corporation	Japan	1	1	4	4	25	25
Con Agra	United States	3	3	10	13	10	23
Cargill Inc	United States	3	2	7	9	43	22
Ajinomoto	Japan	2	2	6	9	33	22
Uniq	Western Europe	1	1	5	5	20	20
Kerry Group PLC	Western Europe	–	1	–	5		20
Snow Brand Milk Products	Japan	1	1	2	5	50	20
Maruha Corp.	Japan	1	1	1	6	100	17
Sudzucker	Western Europe	–	1	–	7	–	14
Meiji Dairies Co.	Japan	–	1	–	9	–	11
Orkla	Western Europe	–	1	–	10	–	10
Scottish & Newcastle	Western Europe	ns	ns	1	2	ns	ns
Grand Metropolitan	Western Europe	2	–	4	–	50	–
Guinness	Western Europe	1	–	3	–	33	–
Northern Foods	Western Europe	2	–	5	7	40	–
Tomkins	Western Europe	2	–	9	7	22	–
Maple Leaf Foods	Rest of the world	1	–	3	6	33	–

[a]SGI = (number of total business activities in overseas markets)/(number of total business activities).

TABLE A1.9. Geographical globalization indicator of leading food MNEs in 1988 and 1999-2000.

MNE	Home zone of the parent company	GDI[a] (%)		IZI[b] (%)		GGI[c] (%)	
		1988	2000	1988	2000	1988	2000
Seagram Co.	Rest of the world	75.0	87.5	85.3	98.3	64.0	86.0
Unilever	Western Europe	75.0	87.5	87.0	94.6	65.3	82.8
Nestle	Western Europe	75.0	87.5	94.4	94.0	70.8	82.3
H. J. Heinz	United States	62.5	87.5	90.7	89.4	56.7	78.2
Danone	Western Europe	50.0	87.5	55.2	89.1	27.6	78.0
Novartis	Western Europe	75.0	87.5	92.3	86.4	69.2	75.6
Cadbury Schweppes	Western Europe	50.0	87.5	76.6	82.9	38.3	72.5
Heineken	Western Europe	75.0	87.5	77.8	82.4	58.3	72.1
Pepsico	United States	62.5	87.5	28.5	77.1	17.8	67.5
WM. Wrigley Jr. Co.	United States	–	75.0	–	89.2	–	66.9
Pernod Ricard	Western Europe	25.0	75.0	45.3	88.3	11.3	66.2
Sara Lee Corporation	United States	5.0	75.0	81.7	85.5	40.9	64.1
Coca-Cola	United States	75.0	87.5	70.3	73.0	52.7	63.9
Montedison (EBS)	Western Europe	37.5	75.0	56.5	83.9	21.2	62.9
Parmalat	Western Europe	–	75.0	–	83.6	–	62.7
Numico	Western Europe	–	75.0	–	81.7	–	61.3
Tate & Lyle	Western Europe	62.5	75.0	70.5	77.9	44.0	58.4
Friesland Dairy Foods	Western Europe	–	75.0	–	75.0	–	56.3

64

Company	Region						
Kellogg Company	United States	62.5	75.0	80.0	74.7	50.0	56.0
Bunge International	Rest of the world	62.5	62.5	83.4	89.0	52.1	55.6
Sodiaal	Western Europe	62.5	75.0	85.5	69.6	53.4	52.2
Philip Morris	United States	62.5	75.0	46.7	69.2	29.2	51.9
Interbrew	Western Europe	12.5	50.0	41.9	95.7	5.2	47.8
Danisco	Western Europe	–	50.0	–	90.7	–	45.3
Barilla	Western Europe	12.5	50.0	75.0	87.5	9.4	43.8
Mc Cormick & Co.	United States	50.0	62.5	57.1	68.6	28.6	42.9
Procter & Gamble	United States	ns	50.0	68.3	84.9	ns	42.5
General Mills	United States	37.5	62.5	71.9	66.7	27.0	41.7
Ferrero	Western Europe	37.5	50.0	82.1	82.4	30.8	41.2
Maruha Corp.	Japan	62.5	62.5	95.5	64.1	59.7	40.0
LVMH	Western Europe	37.5	50.0	72.8	79.2	27.3	39.6
Bongrain	Western Europe	37.5	62.5	7.0	62.8	2.6	39.2
Arla Foods	Western Europe	ns	62.5	12.5	62.5	ns	39.1
Ralston Purina	United States	50.0	50.0	18.2	77.8	9.1	38.9
Mars Inc.	United States	37.5	50.0	77.8	76.5	29.2	38.2
Kerry Group PLC	Western Europe	–	62.5	–	60.0	–	37.5
Nutreco	Western Europe	–	50.0	–	73.0	–	36.5
Nichirei Corp.	Japan	37.5	37.5	61.5	95.7	23.1	35.9
CSM	Western Europe	–	37.5	–	89.7	–	33.6
Carlsberg	Western Europe	37.5	50.0	55.0	67.3	20.6	33.6

TABLE A1.9 (continued)

MNE	Home zone of the parent company	GDI^a (%)		IZI^b (%)		GGI^c (%)	
		1988	2000	1988	2000	1988	2000
Allied Domecq PLC	Western Europe	37.5	37.5	44.0	88.9	16.5	33.3
Nissin Food Products	Japan	12.5	37.5	21.4	79.4	2.7	29.8
Koninklijke Wessanen	Western Europe	12.5	37.5	44.2	79.2	5.5	29.7
Quaker Oats Company	United States	37.5	50.0	69.7	58.3	26.1	29.2
Suntory	Japan	37.5	50.0	83.3	55.3	31.3	27.7
Dole Food	United States	37.5	50.0	44.7	53.8	16.8	26.9
Nippon Suisan	Japan	37.5	37.5	100.0	64.0	37.5	24.0
Oetker Gruppe	Western Europe	–	50.0	–	45.8	–	22.9
Kirin Brewery Co. LTD.	Japan	37.5	62.5	25.6	35.7	9.6	22.3
Kikkoman	Japan	12.5	37.5	83.3	58.6	10.4	22.0
Meiji Seika Kaisha	Japan	37.5	37.5	58.3	58.5	21.9	22.0
Con Agra	United States	12.5	75.0	5.7	29.2	0.7	21.9
Associated British Foods	Western Europe	25.0	50.0	23.2	43.6	5.8	21.8
McCain Foods LTD	Rest of the world	25.0	37.5	52.5	54.8	13.1	20.5
Chiquita Brands	United States	25.0	37.5	29.7	51.1	7.4	19.2
Orkla	Western Europe		37.5	–	46.7	–	17.5
Ajinomoto	Japan	25.0	25.0	86.4	69.6	21.6	17.4
Archer Daniels Midland	United States	25.0	37.5	35.3	46.3	8.8	17.4
Nichiro Gyogyo	Japan	37.5	37.5	100.0	45.5	37.5	17.0
Cargill Inc.	United States	75.0	62.5	52.1	26.2	39.1	16.4

Company	Region						
Land O'Lakes	United States	50.0	62.5	20.8	25.9	10.4	16.2
Campbell Soup	United States	50.0	37.5	45.7	42.1	22.8	15.8
Tyson Foods Inc.	United States	25.0	37.5	23.1	41.7	5.8	15.6
Smithfield Foods	United States	–	37.5	–	41.2	–	15.5
Contigroup	United States	–	37.5	–	40.0	–	15.0
Sudzucker	Western Europe	–	37.5	–	37.0	–	13.9
Hershey Foods	United States	37.5	37.5	46.2	36.4	17.3	13.6
Nippon Meat Packers	Japan	50.0	37.5	80.0	34.9	40.0	13.1
South African Breweries	Rest of the world	–	37.5	–	28.1	–	10.5
Campina Melkunie	Western Europe	37.5	25.0	45.5	40.0	17.0	10.5
Morinaga Milk Industry	Japan	25.0	25.0	66.7	37.5	16.7	9.4
Snow Brand Milk Products	Japan	12.5	37.5	75.0	25.0	9.4	9.4
QP Corporation	Japan	12.5	25.0	15.4	36.8	1.9	9.2
IBP Inc	United States	–	50.0	–	18.2	–	9.1
Yamazaki Baking	Japan	ns	25.0	31.3	35.0	ns	8.8
ITO Ham Foods	Japan	12.5	25.0	35.3	33.3	4.4	8.3
Lactalis	Western Europe	12.5	25.0	40.0	31.4	5.0	7.9
Hormel Foods	United States	12.5	25.0	30.0	31.3	3.8	7.8
Diageo PLC	Western Europe	–	12.5	–	58.3	–	7.3
Uniq	Western Europe	12.5	12.5	19.6	48.3	2.4	6.0
Anheuser Busch Inc.	United States	50.0	25.0	24.1	23.8	12.1	6.0
International Multifoods	United States	12.5	12.5	29.4	46.2	3.7	5.8

TABLE A1.9 (continued)

MNE	Home zone of the parent company	GDI[a] (%)		IZI[b] (%)		GGI[c] (%)	
		1988	2000	1988	2000	1988	2000
Earthgrains Company	United States	–	12.5	–	45.5	–	5.7
Glanbia	Western Europe	–	12.5	–	43.3	–	5.4
Toyo Suisan	Japan	12.5	12.5	50.0	40.0	6.3	5.0
Farmland Industries	United States	–	12.5	–	34.1	–	4.3
Nisshin Flour Milling	Japan	12.5	12.5	33.3	32.0	4.2	4.0
Constellation Brands	United States	–	12.5	–	29.6	–	3.7
Sapporo Breweries	Japan	12.5	12.5	42.9	25.0	5.4	3.1
Suiza Dairy Group	United States	–	25.0	–	10.3	–	2.6
Asahi Breweries	Japan	–	12.5	–	20.0	–	2.5
Dean Foods	United States	25.0	25.0	8.7	5.6	2.2	1.4
Ezaki-Glico	Japan	25.0	12.5	17.6	10.5	4.4	1.3
Meiji Dairies Co.	Japan	–	12.5	–	8.0	–	1.0
Whitbread & Co.	Western Europe	37.5	–	39.0	6.7	14.6	–
Tomkins	Western Europe	37.5	–	38.8	43.7	14.5	–
Maple Leaf Foods	Rest of the world	12.5	–	50.0	11.1	6.3	–
Northern Foods	Western Europe	12.5	–	5.6	10.5	0.7	–

[a]GDI = Share of host zones in world total in total number of geographical zones
[b]IZI = Share of foreign subsidiaries in the total number of subsidiaries
[c]GGI = Geographical globalization indicator = GDI · IZI

NOTES

1. These business activities are classified according to the United Nations' four-digit Standard Industrial Classification. Since the end of the 1980s, the Agrodata research team has been developing the UN SIC using six-digit classes in order to better respond to the high segmentation of food markets.

2. Annual company financial reports are the main source of information. Each company also has a press book that includes all information on the company found in specialized periodicals, magazines, newpapers, (*Agia-Alimentation, RIA, La Tribune, Les Echos, The Financial Times, Fortune International, The Grocer, Tokyo Business Today,* etc.) and international Internet databanks (Hoover's, Fortune Directory, Edgar's Online, Nikkei's, etc.).

3. In Agrodata, a food-processing enterprise is classified in a food sector if it realizes more than 25 percent of its overall sales in this specific sector. If the firm operates in more than one sector with a share that does not exceed 25 percent, then this enterprise is identified as a "multiproduct" firm with a brand portfolio.

4. These data are not exhaustive and are limited by the number and nature of the information resources used by Agrodata. Nevertheless, their aggregate analysis gives important indicators of evolutionary trends of MNEs.

5. Greenfield investments, the entry of foreign capital into a host country with a new production project, require time and effort to build new plants, create new activities, and train unspecialized staff and a workforce amid market uncertainties. These investments become sunk costs if the newly created subsidiary or recently built plant is sold. On the contrary, mergers and acquisitions are, more and more, transactions arranged by stock-exchange markets. The transaction price includes transaction costs which balances out some of the sunk costs.

6. These eight zones are Africa, Latin America, Asia, North America, Western Europe, ECE (Eastern and Central Europe including Russia), the Mediterranean, and Oceania.

7. We are aware of the differences between subsidiaries arising from their size, function, and sales. However, it is impossible to find and to include all this information concerning the different characteristics of each subsidiary for the top 100. So, the results, far from being exhaustive, must be handled with caution. On the other hand, the size or nature of the subsidiaries has little influence on the indicators defined here. It is not the size of the subsidiary that is measured, but the even distribution of the MNE's activities over a number of geographical zones. Recent events (e.g., Tyson Foods took over IBP; the former being more global than the latter, even if it is much smaller in size) seem to corroborate this. Moreover, the following chapters of this book deal with the different features of the MNEs' subsidiairies in detail.

8. GDI = # of the foreign host zones/8 geographical zones of Agrodata; IZI = # of foreign subsidiaries of the food MNE/# of its total subsidiaries; GGI = GDI · IZI.

9. These data are presented in the annual reports of Groupe Danone, Nestlé, PepsiCo, and The Coca-Cola Company.

10. Hillsdown Holdings, Rank Hovis Mc Dougall, Associated British Foods.

11. We can give the examples of the Dutch giants mentioned previously, the Italian dairy processor Parmalat, French multinationals such as Pernod Ricard or Bongrain, and brewers originating in Belgium, Holland, and Denmark.

12. North America, Japan, Western Europe, Australia, and New Zealand.

13. Calculations based on data presented in Tables A1.1, A1.2, and A1.3.

14. These business activities (sectors) correspond to the UN's four-digit SIC classification, developed by the Agrodata research team on a six-digit basis since the mid-1980s, in order to take into account the important market segmentation of the food-processing sector.

15. SGI = # of business activities in host zone(s) of the food MNE / # of business activities in its home zone

16. Of course, these indicators must not be taken as absolute measurements of a taxonomy of MNEs and must be coupled with more in-depth analysis of the growth strategies of these food-processing multinationals, mostly based on case studies.

17. Refocusing on core businesses is discussed in detail, with the application of an econometric analysis, elsewhere in this book.

18. Agway, Coberco, Gold Kist, Groupe Saint Louis.

19. Takeovers include Pillsbury by General Mills, Cerestar by Cargill Inc., Nabisco by Philip Morris, Quaker Oats by PepsiCo, IBP by Tyson Foods.

20. This group comprises MNEs originating in Canada, Australia, South Africa, Latin America, and Far East Asia.

BIBLIOGRAPHY

Aaker, David A. and Joachimsthaler, E. (1999). The lure of global branding. *Harvard Business Review,* 77: 137-144.

Braudel, F. and Duby, G. (Eds.) (1985). *La Méditerranée: L'espace et l'histoire.* Paris: Flammarion.

Dyer, Jeffrey H. (1997). Effective interfirm collaboration: How firms minimize transaction costs and maximize transaction value. *Strategic Management Journal,* 18: 535-556.

Foss, Nicolai J. and Iversen, Mikael (1997). *Promotion synergies in multiproduct firms: Toward a resource-based view.* Department of Industrial Economics and Strategy, Copenhagen Business School, Working Paper no. 97-12. Copenhagen.

Garrette, Bernard (1989). Actifs spécifiques et coopération: Une analyse des stratégies d'alliance. *Revue d'Economie Industrielle,* 50(4): 15-31.

Ghemawat, Pankaj and Ghadar, Fariborz (2000). The dubious logic of global megamergers. *Harvard Business Review,* 78: 66-72.

Henderson, Dennis R., Handy, Charles R., and Neff, Steven A. (1996). *Globalisation of the processed foods market.* Food and Consumer Economics Division, Economic Research Service, Agricultural Economic Report No. 742, USDA. Washington, DC: GPO.

Levitt, Theodore (1983). The globalization of markets. *Harvard Business Review,* 61: 1-11.

Malassis, Louis (1973). *Economie agro-alimentaire.* Volume 1, *Economie de la consommation et de la production agro-alimentaire.* Paris: Cujas.

Markusen, James R. (1998). Multinational enterprises and trade theory. In Pick, Daniel H., Dennis R. Henderson, Jean D. Kinsey, and Ian M. Sheldon (Eds.), *Global markets for processed foods: Theoretical and practical issues* (pp. 95-120). Boulder, CO: Westviews Press.

Mendez, Ariel (1997). Multinationalisation: La dynamique organisationnelle. *Revue Française de Gestion,* 6: 18-24.

Mucchielli, Jean-Louis (1998). *Multinationales et mondialisation.* Paris: Seuil.

Narula, Rajneesh and Dunning, John H. (1999). Developing countries versus multinationals in a globalizing world. *Forum for Development Studies,* 2: 261-287.

Oman, Charles, Chesnais, François, Pelzman, Joseph, and Rama, Ruth (1989). *Les nouvelles formes d'investissement dans les industries des pays en développement.* Paris: OCDE, Etudes du Centre de Développement.

Palpacuer, Florence (1999). Competence-based strategies and global production networks: A discussion of current trend changes and their implications for employment. *Competition and Change: The Journal of Business and Political Economy,* 4(4): 1-48.

Pérez, R. and Palpacuer, Florence (Eds.) (2002). *Mutations des modes de gouvernance, dynamique de compétitivité et management stratégique des firmes: Le cas des firmes multinationales alimentaires en Europe.* ERFI et GRAAL, Commissariat Général du Plan-Services des Etudes et de la Recherche, Appel d'offres sur Gouvernement d'entreprise. Décision n°7/1999 du 09/08/99. Montpellier.

Petrella, Ricardo (1990). La mondialisation de l'economie par la compétitivité. *Politique Economique,* 46: 21-33.

Porter, Michael (1986). Conceptual foundations. In Porter, Michael (Ed.), *Competition in global industries* (pp. 15-60). Boston, MA: Harvard Business School.

Porter, Michael (1991). Toward a dynamic theory of strategy. *Strategic Management Journal,* 12: 5-117.

Rastoin, Jean-Louis, Ghersi, Gérard, Pérez, Roland, and Tozanli, Selma (1998). *Structure, performance et stratégies des grands groupes agro-alimentaires multinationaux.* Montpellier, France: CIHEAM-IAMM.

Rugman, Alan M. (1998). Foreign direct investment and internalization in processed foods. In Pick, Daniel H., Dennis R. Henderson, Jean D. Kinsey, and Ian M. Sheldon (Eds), *Global markets for processed foods: Theoretical and practical issues* (pp. 121-134). Boulder, CO: Westviews Press.

Teece, David T., Pisano, G., and Shuen, A. (1997). Dynamic capabilities and strategic management. *Strategic Management Journal,* 18(7): 509-533.

Tozanli, Selma (1998). Capital concentration among the food multinational enterprises and development of the world's agro-food system. *International Journal of Technology Management,* 16(7): 695-703.

Tozanli, Selma (2000). Competing in a global economy: Some reflections on how the food processing TNCs restructure their value chains. Discussion paper presented at the UNCTAD Discussion Forum "What Can Value Chain Analysis Reveal About the Unequal Distribution of the Gains from Globalisation?" Geneva, November 20.

Tozanli, Selma (2001). Structural changes in the French dairy processing industry during the last two decades. *Bulletin of the International Dairy Federation,* 36: 30-42.

Williamson, Oliver E. (1991). Strategizing, economizing, and economic organization. *Strategic Management Journal,* 12: 75-94.

Williamson, Oliver E. (1996). *The mechanisms of governance.* New York: Oxford University Press.

Williamson, Oliver E. (1999). Strategy research: Governance and competence perspectives. *Strategic Management Journal,* 20: 1089-1108.

Zander, Ivo (1998). The evolution of technological capabilities in the multinational corporation: Dispersion, duplication, and potential advantages of multi-nationality. *Research Policy,* 27: 17-35.

Chapter 2

The Performance of Multinational Agribusinesses: Effects of Product and Geographical Diversification

George Anastassopoulos
Ruth Rama

INTRODUCTION

In this chapter, we will study the impact of geographical and industrial diversification on the growth rates of the world's 100 largest food and drink multinationals (the top group). Among the large firms that influence the food and beverage (F&B) industry worldwide, growth has been an important issue, possibly even more important than for multinationals in other industries. These firms account for one-third of production (Rastoin et al., 1998) and more than one-half the technological activities of the world food and beverage industry (see Chapter 3). Food and beverage multinationals (FBMs) have often aimed at gaining size. They perceived that large companies had the opportunity to enter new, profitable markets for foodstuffs (sauces, precooked food, etc.) while keeping oligopolistic control over markets for basic foodstuffs (sugar, flour, etc.), where margins are thinner. Small companies often lack this possibility because they cannot balance thin margins with the enormous volumes of foodstuffs marketed by the largest firms. The firms analyzed in this chapter grow at an average annual rate 3 to 4 percent higher than that of global F&B production (Rastoin et al., 1998). However, they have been losing dy-

The authors would like to thank Frangkiskos Filippaios for research assistance.

namism over recent years. The exceptions are large companies from Japan and some European countries, which are expanding swiftly. For rival firms, this could also be a matter of concern.

Analyzing the effect on FBM growth of diversification could help to fill a gap in international business studies, since their specificities could be concealed in cross-sectional analyses that would include all kinds of multinationals. For instance, country spread could be a key aspect of geographic diversification particularly important in the route to expansion of FBMs. Proximity to the market is crucial for multinationals in consumer products (Ietto-Gillies, 2002). On the other hand, FBMs serve foreign markets by foreign production rather than by exports (Pearce, 1993). Finally, unlike international manufacturers of cars or computers, FBMs interact with varied national consumers. Although diets tend toward homogeneity, especially in North America and Western Europe (Connor, 1997), peculiarities of national consumptions remain substantial (Christensen, Rama, and von Tunzelmann, 1996).

Another characteristic of FBMs is that they could be considered precursors to other multinationals because they are, on average, older and internationally more experienced (Stopford and Dunning, 1983). Thus, the analysis of FBMs helps toward a better understanding of the rationale of other international businesses.

The longevity of FBMs seems to be associated with their strategies of growth. Leading FBMs have sometimes survived for more than a century thanks, among other reasons, to their strategies of diversification into a variety of industrial and geographic markets. By spreading into new industries in the domestic market and investing abroad, large U.S. producers of branded and packaged foods, such as Borden, Heinz, Campbell Soup, or Del Monte, grew continuously throughout the first decades of the twentieth century (Chandler, 1990). Food leaders diversified their lines of products chiefly to exploit economies of scope in distribution; many also invested in retailing outlets. Their strategies were very effective. Unlike very large U.S. and British F&B companies, most of the giant firms that dominated the German F&B industry by the end of the nineteenth century did not survive, for reasons that included the failure of German enterprises to diversify in time into either new products or new geographic markets. Even today, very few German companies are among the world's 100 largest FBMs.[1] Leading U.S. food manufacturers went on diversifying into new product markets from 1919 to 1972, but this trend stabilized toward the end of the 1970s (McDonald, Rayner, and Bates, 1989).

During the 1980s, doubts about the current effectiveness of unrelated diversification started to be cast. Researchers suggested that related diversification was superior to unrelated diversification because the latter discourages synergies among activities (Palich, Cardinal, and Miller, 2000; Palich, Carini, and Seaman, 2000). Knowing which type of industrial diversification is more conducive to good performance is important, since many large F&B firms are conglomerates (unrelated companies), a characteristic of the industry that has increased since the 1980s, when several giant companies from other sectors, especially the tobacco industry, diversified into F&B (Rama, 1992). Since the 1980s, analysts also began encouraging F&B firms to focus on core businesses and divest from noncore businesses (Ding and Caswell, 1995).

This advice was followed by most large F&B enterprises. In a sample of ten large U.S. food-manufacturing companies, only five reduced their unrelated diversification over 1981-1989 and focused, as recommended, on related diversification (Ding and Caswell, 1995). Over the 1980s and early 1990s, however, the world's largest F&B companies invested billions of dollars in mergers and acquisitions (M&A) (see Chapter 1), chiefly because they wanted to purchase new concerns in their trade and divest from unrelated, often unprofitable businesses. Among the world's 100 largest FBMs (hereafter, the top group), sales of nonfood products as a proportion of total sales fell from 25.7 to 11.2 percent from 1981 to 1988, a decline suggesting increased concentration in core businesses (Rama, 1998). We do not know, however, if such a strategy actually stimulated growth, because rates of growth by FBMs tended to decline over the 1990s.

The geographic diversification of FBMs poses additional questions. Though anecdotal evidence suggests that the companies have expanded quickly worldwide over recent periods, we do not know exactly where such firms have been expanding and whether companies from different countries display different geographic strategies.

Even the effectiveness of such a strategy is now controversial. Rugman (1996) has argued that multinational firms can reduce their profit risk by engaging in foreign operations. Some researchers based on Penrose's (1959) work have suggested, however, that the burden of organizing networks all over the world could harm the performance of the highly international company. In an era of global expansion of FBMs, it is useful to explore this topic in detail. Some important aspects have been analyzed in Chapter 1. However, multinationality has many other facets; their influence on performance will be studied here. The number of countries into which a firm expands, the share of foreign affiliates in total affiliates, or the number of affiliates by host country could each affect the rate of growth in different and even antagonistic ways.

Here we investigate firms belonging to a variety of home countries as well as industries; to analyze them, we use panel data covering 1985 to 1996. The next section reviews the theoretical background that informs this investigation. The subsequent sections present the data and the empirical analysis, a discussion of our results, and some conclusions.

THEORETICAL BACKGROUND

Because the number of studies dealing with industrial and geographical diversification in FBMs is limited, we also supplement our research with studies covering nonfood multinationals and single-country firms.

Industrial Diversification

Over the past three decades, researchers have devoted an enormous amount of attention to industrial diversification, i.e., expansion into product markets new to a firm (Hitt, Hoskisson, and Kim, 1997). The topic has become "the dominant research stream in the field of strategic management" (Sambharya, 2000, p. 163).

Authors have used different tools to measure diversification and have consequently pursued different inquiries. The authors who have measured the number of industries into which firms have diversified (the quantitative approach) have chiefly investigated whether diversified firms were more profitable than single-business companies. In general, the response has been that diversified firms enjoy more advantages (for instance, economies of scope) than single-business companies (Palich, Cardinal, and Miller, 2000). The researchers who have used a qualitative approach, based on the underlying expansion logic behind the company behavior (Sambharya, 2000), have posed a different question: Does the diversified-related enterprise perform better than the unrelated enterprise (conglomerate)? Rooted in the resource-based theory of the firm (Penrose, 1959), some authors have replied that the related firm performs better because it can minimize risks, cross subsidize different lines of businesses, spread the prestige of its brands into a variety of products, distribute the costs of already capitalized assets, or exploit interrelationships between different technologies and marketing channels (for a survey of studies on industrial diversification, see Palich, Cardinal, and Miller, 2000; for critical views of theories of diversification based on economies of scope and transaction-cost economics, see Dosi, Teece, and Winter, 1992). Other authors, however, have noticed that in practice synergies are difficult to achieve because related firms need great financial resources and managerial ability to exploit interrelationships. Markides and Williamson (1994) point, moreover, to the firm needing strategic relatedness (brands, distribution, technology) in addition to market relatedness in order to obtain full benefits from its diversification strategy. Dass (2000) suggests that abrupt changes in product diversity could have negative effects on performance. On the other hand, as some authors claim, diversification into nonrelated activities can reduce a firm's ability to grow since it increases the complexity of the network (Williamson, 1975). Others argue, though no evidence is yet provided, that the important question is whether lines of business, even if not related within the same two-digit industry, display coherence, i.e., common technological and marketing characteristics (Dosi, Teece, and Winter, 1992).

Unfortunately, studies devoted to industrial diversification and performance in multinationals are not conclusive. The highly international firm often develops complex structures which could constrain its growth because they involve large organizational and management costs. Analyzing 200 U.S. and European multinationals in all industries over 1977-1981, Geringer, Beamish, and daCosta (1989) remark that related companies tended to perform better than unrelated ones. Among Japanese multinationals analyzed over 1977-1993, this could also be the case, but it was not always so (Geringer, Tallman, and Olsern, 2000).

The situation in FBMs is little known because analyses on this topic are rare. A study of sixty-four major FBMs finds that related companies, highly diversified within the food and beverage industry, grew faster than other FBMs over 1977-1988 (Rama, 1998). However, the study argues that related companies did not necessarily outperform their rivals over the early 1990s, probably because concentration in core businesses had become by then a fairly extensive strategy in the international agrofood sector. As mentioned, marketing and technological coherence are important. Analyzing 792 very large enterprises from a variety of industries, another study notices that in 1982 food companies were less diversified than other manufacturing firms (Pearce, 1993). The author argues that one reason why food, drink, and tobacco firms spread within these industries as well as into chemicals or pharmaceuticals is that such products are often marketed through similar channels (supermarkets, drugstores, etc.). This circumstance, he adds, contributed to economies of scope. In our view, the FBM holding chemicals businesses could enjoy not only marketing but also technological economies of scope because of often developing innovative activities in chemicals (see Chapter 3).

Over time a consensus has tended to emerge in empirical studies. Palich, Cardinal, and Miller (2000), who surveyed and synthesized eighty-two quantitative studies on the diversification-performance linkage, conclude that most empirical results support the curvilinear model: diversified firms tend to outperform less-diversified ones, although performance declines as companies venture into a greater number of unrelated industries.

Diversification into upstream and downstream industries such as agriculture or retailing (vertical relatedness) poses other questions. Though multinational agribusinesses often diversified into agriculture during their early expansion (Dunning, 1993), traditional vertical relatedness seems to no longer be a source of good performance. In the top group of FBMs, the companies that held farms (in addition to food and drink factories) were not necessarily fast growers over 1977-1994; those holding retail outlets grew above average in the 1970s and 1980s but not in the 1990s, when relationships between food manufacturers and retailers became less antagonistic (Rama, 1998). This result coincides with that of Fan and Lang (2000) who find, in a sample of around 5,000 firms from all manufacturing sectors, that vertical relatedness is associated with poor performance. However, when a firm diversifies into knowledge-creating activities it can utilize its experience and resources in ways leading to lowering the costs of expansion (Cantwell, 1989; Cantwell and Hodson, 1991).

In a sample of sixty-four large FBMs, those that had diversified into biotechnology, veterinary services, microbiological products, and other activities high in technological content grew swiftly over 1977-1988, allowing for size, multinationality, profit, and other types of industrial diversification (Rama, 1998).

Here we test whether vertical relatedness helps FBMs to expand. We explore the effects of diversification into agriculture, retailing, and high-tech activities (checking for other factors of growth).

Geographical Diversification

Researchers do not agree regarding the effects on performance of geographical diversification, i.e., expansion of the company into different geographical markets. First, it is not clear if highly international companies are fast growers. Some authors find that the effects of multinationality on performance are very weak, if any (Tallman and Lee, 1996). When comparing 143 multinationals and 128 very large domestic firms of various sectors, Cantwell and Sanna-Randaccio (1993) observe, moreover, that the companies that had remained domestic grew faster. Siddarthan and Lall (1982), who studied the largest U.S. multinationals at the end of the 1970s, found that multinationality exerts a negative influence on growth. Second, some authors argue that the association between multinationality and performance is not linear. Among 200 large U.S. and European multinationals, Geringer, Beamish, and daCosta (1989) find that, whatever their continent of origin, highly international companies perform better. However, after a point, performance peaks and tends to diminish. According to the authors, the costs associated with geographic dispersion could escalate after some "internationalization threshold." Finally, authors of other studies notice that the effects of multinationality on performance seem to change both across and within industries. In their subsample of 143 large multinationals, Cantwell and Sanna-Randaccio (1993) observe two different statistical results, depending on the characteristics of the industry to which the company belongs. Only in global, i.e., internationally integrated, industries does the highly international company grow faster than the less international one (controlling for size, technology, and market power).[2] By contrast, in multidomestic industries (non–internationally integrated industries), highly international firms are not necessarily fast performers. The authors explain that, in such industries, the highly international firm endures the costs of a complex organization without benefiting from the integrated networks needed to exploit locational advantages of supply and demand; thus it performs less well. Since the F&B industry is considered multidomestic (not global), we could infer that highly international food firms tend to grow slowly, though Cantwell and Sanna-Randaccio's results are not disaggregated by industry. A sectoral study shows, however, that some highly international food firms are actually fast growers (Rama, 1998). Effects of multinationality are negative in old, experienced FBMs but not in younger FBMs. Old and experienced FBMs are more likely than younger, less experienced ones to have moved from old (North America, Western Europe) to new geographic national markets, which are often deprived of adequate retailing facilities or initially reluctant to adopt a foreign diet. Owing to such handicaps, the FBM could have endured some periods of restrained growth.

Here we test whether highly international FBMs grow faster (other determinants of growth also checked for). Given that previous findings could diverge because authors use different measures highlighting different angles of the question, we test the hypothesis with a variety of proxies of geographical diversification.

DIVERSIFICATION, SIZE, AND RATE OF GROWTH

This section provides an insight into the diversification, size, and growth of FBMs.

The Data

Our sample of FBMs comprises 100 companies included in the top group over the late 1980s to 1996 (see Appendix). The source of the information is AGRODATA (Institut Agronomique Méditerranéen de Montpellier [IAMM], 1990; Padilla et al., 1983; Rastoin et al., 1998). Produced by IAMM (France), this database has gathered information on the world's 100 largest food multinationals since the 1970s. The firms in our sample are active in a variety of industries, including meat processing, confectionery, dairy products, canned specialties, and spirits. All are food or beverage processors, and a number of them also have agribusinesses and other concerns. The sources for AGRODATA are *Moody's Industrial Manual,* the Fortune Directory of the 500 largest U.S. and the 500 largest non-U.S. corporations, the "Dossier 5,000" of the largest European companies published by *Le Nouvel Economiste,* Dun & Bradstreet, and the annual reports of the enterprises, among others.

Researchers measure industrial diversification in multiple ways, each with its own merits and shortcomings (for surveys, see Davis and Duhame, 1995; Fan and Lang, 2000; Sambharya, 2000). Here we test whether diversified FBMs, both related and unrelated, grow quickly (after allowing for size, financial variables, and geographical diversification). In addition to counting the four-digit SIC categories in which the FBM operates, as studies in the *quantitative* stream do, we measure the economic weight of related activities, i.e., the food sales/total sales ratio at the company level. Because the advantage of the related firm is in business synergy, the economic *dimension* of related networks (not only their complexity) is important. Researchers also measure geographical diversification using different proxies, which throw light on different angles of the picture (for a review, see Dörrenbächer, 2000). Some authors use more than one variable in the same equation since they recognize that different aspects of internationalization could have different, and even antagonistic, results on company growth. For instance, Cantwell and Sanna-Randaccio (1993) use six variables measuring internationalization in their model to better specify the route to growth in global industries. In our equation, we use three: country spread; the ratio of foreign affiliates to total number of affiliates; and ratio of foreign affiliates to total number of host countries. The three variables measure different aspects of internationalization. Country spread, which reflects the number of geographic markets in which the company is active, seems, as stated previously, especially important for FBMs. The foreign affiliates/total affiliates ratio measures the international projection of the company and the relative emphasis on foreign versus domestic activities. It is comparable to the foreign sales/total sales ratio (Ietto-Gillies, 2002).[3] The foreign affiliates/total host-countries ratio measures the complexity of networks at the host-country level. Organizing different businesses within the same host country, for instance, is considered to be a complex task (Prahalad and Doz, 1987). This situation could occur when multiproduct companies, such as those studied here, establish many affiliates by country.

Diversification Patterns in Food and Beverage Multinationals

The purpose of this section is twofold. First, we describe the pattern of industrial and geographical diversification of the world's 100 largest FBMs. In doing so, we provide a description of the sample in both aspects of diversification, as well as other basic characteristics of the firms (i.e., size and growth rates) over 1985 to 1989 and 1990 to 1996. We calculate averages of the variables for groups of companies and by company for both periods.

Time averages for individual firms are calculated as follows:

$$\overline{V}_i = \frac{1}{T}\sum_{t=1}^{T} V_{it} \qquad (2.1)$$

where i represents the company, t is the time horizon, and V is one of the variables considered in this chapter.

Group averages for firms with common characteristics are calculated in this way:

$$\overline{\overline{V}} = \frac{1}{N}\sum_{i-1}^{N} \overline{V}_i \qquad (2.2)$$

where again i represents the company.

Here we use the time average for individual firms (see Equation 2.1) to construct group averages for companies displaying common characteristics, i.e., FBMs from the same home country or home region. This enables us to obtain means and statistics across different time periods and different companies at the same time. It is a straightforward way of aggregation over different firms and time periods. Second, as mentioned, we empirically explore the linkages between variables measuring the growth of a company and its diversification status.

Using data on the affiliates, domestic and foreign, Table 2.1 shows the industrial and geographic diversification of the firms. Data refer to the percentage of affiliates each firm possesses in each geographical region and industrial sector. Equivalent data concerning sales are unavailable. On the horizontal axis, the first column reports the home region of the firm; the second names the host region. The vertical axis represents the industrial activity of the affiliate. The categorization is based on Rama (1998), who used the United Nations-Standard Industrial Classification of the affiliates provided by AGRODATA (see Table 2.7).

Most FBMs Locate in the Home Region and Diversify Within Core Businesses

Geographical diversification varies with the origin of the FBM, though, on average, the majority of the affiliates locate in the parent company's home region. The percentage of affiliates located in the home region to the total number of affiliates ranges from 42.4 percent for North American (United States and Canada) multinationals, which seem to be overall the most geographically diversified, to 70.3 percent for European companies, the least geographically diversified (see Table 2.1). An F-test confirms that, as already mentioned, home-country and geographic strategies are statistically associated (F = 5980.7; p = 0.01).

Regarding industrial diversification, the majority of affiliates, as expected, are gathered within the core sector of the firm, i.e., F&B (see Table 2.1). The exceptions are again North American firms, which tend to diversify into noncore businesses: 46.4 percent of their affiliates are in other activities, i.e., nonfood. By contrast, the most focused on core businesses are the European FBMs. Companies from different home countries tend to accord different relative importance to different businesses. Asian (mostly Japanese) FBMs give special importance to their technologically related affiliates, the majority also located in Asia. The percentage of North American affiliates specializing in technology is below that in European and, especially, Japanese FBMs. This does not necessarily mean that North American FBMs devote less attention to innovative activities than their foreign rivals. It may well be that multinationals based in different countries organize their R&D activities in a different manner and that North American companies develop theirs in laboratories within manufacturing plants, rather than in specialized affiliates as the Japanese FBMs do. However, the small diversification into specialized laboratories of North American FBMs could be related to the downward trend in the number of their patents, a proxy for innovative activities, which is analyzed in Chapter 3. On the other hand, Japanese and

TABLE 2.1. Industrial and geographical diversification of the world's 100 largest food and beverage multinationals (% of affiliates).

Home region	Host region	Agriculture[a]	Retail[b]	Technology[c]	Within core[d]	Other[e]	Total
Asia	Asia	2.54	11.71	3.09	34.48	17.79	69.61
	Europe	1.33	2.87	0.44	3.31	2.43	10.39
	North America	0.44	1.66	0.55	6.30	2.43	11.38
	ROW	1.55	1.66	0.33	3.43	1.66	8.62
	Total	5.86	17.90	4.42	47.51	24.31	100.00
Europe	Asia	0.06	2.03	0.06	3.97	2.06	8.18
	Europe	0.86	9.41	1.05	41.25	17.77	70.34
	North America	0.37	2.18	0.09	4.21	2.27	9.13
	ROW	0.22	1.66	0.09	8.21	2.18	12.36
	Total	1.51	15.28	1.29	57.64	24.29	100.00

North America	Asia	0.10	1.18	0.03	2.69	3.22	7.22
	Europe	0.10	3.09	0.16	15.04	15.53	33.92
	North America	1.84	4.66	0.43	15.04	20.39	42.36
	ROW	0.39	0.95	0.10	7.75	7.29	16.49
	Total	2.43	9.89	0.72	40.53	46.44	100.00
ROWf	Asia	1.04	1.04	0.00	4.69	1.30	8.07
	Europe	1.56	5.21	0.00	13.02	6.77	26.56
	North America	0.26	1.82	0.00	1.82	1.04	4.95
	ROW	3.91	4.43	0.26	30.47	21.35	60.42
	Total	6.77	12.50	0.26	50.00	30.47	100.00

Source: Authors' calculations based on AGRODATA.

aUN-SIC Codes: 1110, 1210, 1300, 1301, 1302.

bUN-SIC Codes: 6210, 6220, 6300, 6310.

cUN-SIC Codes: 311280, 832020, 832021, 832030, 9320, 9330. The six-digit classification was introduced by AGRODATA to distinguish different activities within four-digit classes.

dForeign affiliates in the same core sector as the mother company.

eForeign affiliates that are nonrelated diversified (nonfood).

fRest of the world (ROW) contains firms based in South Africa, Australia, and South America.

European FBMs diversify much more into retailing than other FBMs. Finally, firms based in Asia and the rest of the world often spread into agricultural businesses. Japanese FBMs invest heavily in fisheries and other primary activities aimed at supplying the home country with food and agricultural products through exports. As will be seen in Chapter 7, Australian and New Zealand companies are also involved in agricultural concerns. A chi-square test shows that home country and behavior of the firm in regard to industrial diversification are statistically associated ($\chi^2 = 17918.2$; $p = 0.01$).

Combining both forms of diversification, we can analyze the geographic distribution of specific businesses within the multinational. Table 2.1 shows a picture of the relative importance of host countries to FBMs in each business. In their F&B businesses ("Within Core" column in the table), North American multinationals operate similar percentages of affiliates in Europe and in North America; the same occurs with retailing businesses. Their affiliates in nonfood businesses ("Other" column) settle in the home region, but location in Europe is also substantial. North American FBMs are an exception; all the other FBMs establish the majority of their affiliates in the same region as the parent and within core activities.

The Economic Weight of Core Businesses
Has Remained Stable

As previously noted, the F&B sales/total sales ratio is used here as a proxy for the economic weight of related activities in the company. Table 2.2 presents the percentage of F&B sales in total sales for two time periods and the major home countries/regions. Two points emerge. First, as shown by a difference of means test, patterns of F&B sales/total sales remained stable ($t = 0.151$; $p > 0.10$). In general, FBMs did not intensify, over the study period, the strategy of sticking to core businesses that started at the beginning of the 1980s (Rama, 1998). Second, an exception is given by Japanese FBMs, since their F&B sales/total sales ratio increased from 1985-1989 to 1990-1996 ($t = 1.522$; $p = 0.10$). In addition, Japanese firms displayed the largest ratio in the sample (94.5 percent in 1990 to 1996) and the smallest standard deviation from the mean (9.2 percent), suggesting that such companies are strongly committed to their main core business.

TABLE 2.2. Industrial diversification of the world's 100 largest food and beverage multinationals, by home country, in percentages.

Home country	Food/total sales			
	Mean	SD	Min	Max
	1990-1996			
United States	88.80	21.51	10.75	100.00
Europe	89.14	19.54	13.50	100.00
United Kingdom	89.85	14.59	32.20	100.00
France	90.44	19.79	34.68	100.00
Rest of Europe	88.05	23.05	13.50	100.00
Japan	94.66	9.19	59.84	100.00
Rest of world	84.31	22.92	38.34	100.00
Total	89.85	18.93	10.75	100.00
	1985-1989			
United States	87.97	19.79	14.16	100.00
Europe	90.05	17.07	17.27	100.00
United Kingdom	90.61	10.34	66.92	100.00
France	91.08	15.68	52.08	100.00
Rest of Europe	88.74	24.20	17.27	100.00
Japan	92.57	12.38	39.35	100.00
Rest of world	85.93	20.68	41.52	100.00
Total	89.68	17.39	14.16	100.00

Source: Authors' calculations based on AGRODATA.

The Pattern of Geographic Diversification Experienced Some Changes

As already mentioned, we will use three different measures of geographical diversification: the total number of foreign countries in which the firms have established foreign affiliates; the ratio of foreign to total affiliates; and the ratio of foreign affiliates to foreign countries (see Table 2.3). The combination of these measures will give us a global picture of the geographical diversification in FBMs.

TABLE 2.3. Geographical diversification of the world's 100 largest food and beverage multinationals, by home country.

	Foreign countries				Foreign/total affiliates (%)				Foreign affiliates/foreign countries			
	Mean	SD	Min	Max	Mean	SD	Min	Max	Mean	SD	Min	Max
1990-1996												
United States	23	20	1	75	51.81	24.99	3.70	90.77	2.06	1.18	1.00	6.29
Europe	20	18	1	98	58.98	21.95	8.11	100.00	2.43	1.28	1.00	6.86
United Kingdom	13	11	1	38	50.37	24.29	8.11	86.67	1.86	0.68	1.00	4.12
France	24	11	5	39	64.04	17.01	35.14	100.00	3.59	1.89	1.30	6.86
Rest of Europe	25	23	4	98	64.53	18.95	20.59	93.64	2.50	1.11	1.00	4.83
Japan	9	6	1	27	47.95	24.30	8.05	93.88	1.96	0.92	1.00	4.10
Rest of world	16	13	2	44	52.42	24.79	21.74	88.97	3.35	1.67	1.20	7.80
Total	18	17	1	98	53.98	23.95	3.70	100.00	2.30	1.27	1.00	7.80
1985-1989												
United States	17	13	2	52	53.40	23.22	7.14	90.48	1.80	1.02	1.00	5.84
Europe	16	15	1	64	53.94	24.31	8.11	100.00	2.48	1.20	1.00	6.86
United Kingdom	10	9	1	28	42.68	22.87	8.11	84.15	2.19	0.77	1.25	3.95
France	18	11	5	39	64.68	21.00	40.00	100.00	3.43	1.72	1.30	6.86
Rest of Europe	23	20	4	64	63.24	21.42	20.59	93.64	2.38	1.13	1.00	4.23
Japan	6	5	1	18	51.14	27.72	8.05	90.48	1.82	1.09	1.00	4.10
Rest of world	15	11	2	31	48.87	24.88	21.74	85.29	3.65	2.18	1.20	7.80
Total	13	13	1	64	52.74	24.82	7.14	100.00	2.21	1.33	1.00	7.80

Source: Authors' calculations based on AGRODATA.

The first part of Table 2.3 reports the average number of foreign countries in which the firm has established operations through a foreign affiliate. The sample contains firms operating in just one foreign country as well as companies operating in ninety-eight different nations by 1996.[4] The evolution of this variable suggests that enterprises from most home countries have become increasingly international over the period analyzed in this chapter. A difference of means test indicates significant differences between 1985 to 1989 and 1990 to 1996 in the country spread of FBMs. The sole exception is provided by FBMs based in the rest of Europe and rest of the world, since the average number of countries in which they disseminated remained unchanged (see Appendix).

By contrast, as shown by a difference of means test, the ratio of foreign to total affiliates remained on average stable ($t = 0.832$; $p > 0.10$). From 1985-1989 to 1990-1996, differences in this ratio were apparent only for British FBMs ($t = 2.199$; $p = 0.01$). This result is similar to that of Ietto-Gillies (2002) for British multinationals in mining and manufacturing, which also increased their international projection. French firms and FBMs based in the rest of Europe were the only ones to settle most of their affiliates in foreign countries over both periods.[5]

Our last measure of diversification is the ratio of established foreign affiliates to the number of foreign countries in which the firm operates. As shown by a difference of means test, from the 1980s to the 1990s, the pattern does not change a great deal ($t = .133$; $p > 0.10$). There are two exceptions. American FBMs significantly increased the number of affiliates they have by country ($t = 2.062$; $p = 0.01$) and British FBMs significantly reduced theirs ($t = -3.113$; $p = 0.01$). French and Japanese firms seem to be the only ones to establish, on average, more than three foreign affiliates in each host country.

In short, although differences between home countries are apparent, the average FBM increased, over the period, its geographic spread. Simultaneously, it maintained the same emphasis on foreign networks versus domestic ones and kept the same organization at the host-country level. It preferred to spread to many nations rather than deepen its already established positions within countries.

U.S. Food and Beverage Multinationals
Are the Largest in Our Sample

This section analyzes the size of FBMs. Though in the econometric analysis we will use size as measured by employment, here we will use two measures of firm size, total assets and employment, in order to show that the characteristics of size are similar whatever the proxy we use. The variables are, respectively, the total assets of the company measured in current U.S.$[6] and the global employment measured as the total number of employees. We present the statistics for both in Table 2.4.

U.S. companies are the largest in our sample whether size is measured in assets or in employment. However, this information should be balanced against that provided by indicators of dispersion, since U.S. multinationals also show the largest standard deviation (SD) of the asset variable. When we measure size by the employment, U.S. firms continue to hold the first position over both periods; Japanese companies remain the smallest. Japanese multinationals are only one-eighth of the size of the average U.S. firm, as measured by assets, and one-half as measured by employment.

Food and Beverage Multinationals Lose Dynamism

As mentioned in the introduction, when FBMs reduced their sales of nonfood products, which was considered a strategy enhancing performance, their rates of growth also tended to fall. In the econometric analysis we will use the growth of sales as the dependent variable. In this section, however, we will use three different measures to capture firm growth over 1985-1996[7]: growth in terms of total assets, employment, and sales (see Table 2.5). We will show that the results, as with size, are similar whatever the indicator.

Whatever the indicator used, on average, FBMs grew much more slowly during the 1990s than over 1985-1989.[8] No matter what measure we use, the growth leaders are French firms and companies from other European countries, which grew almost twice as quickly as multinationals based in other regions.

TABLE 2.4. Descriptive statistics of total assets[a] and employment[b].

	Total assets				Employment			
	Mean	SD	Min	Max	Mean	SD	Min	Max
1990-1996								
United States	8901.421	11271.09	748.00	54917.00	60,509	79,304	5,185	486,000
Europe	6219.057	7564.993	577.08	47001.87	38,438	56,177	1,157	308,000
United Kingdom	5742.798	4394.399	1129.11	17717.37	41,683	22,321	14,573	122,178
France	6864.957	5329.542	1194.66	19165.49	20,546	19,586	6,241	81,579
Rest of Europe	6420.128	10087.51	577.08	47001.87	42,228	80,489	1,157	308,000
Japan	4689.641	3852.307	1183.208	18297.76	7,069	5,687	1,020	22,404
Rest of world	4816.015	4914.912	1003.02	21628.00	46,808	33,776	9,500	109,800
Total	6582.256	8288.794	577.08	54917.00	43,211	63,298	1,020	486,000
1985-1989								
United States	5187.913	6753.811	322.00	38528.00	48,839	47,155	5,002	266,000
Europe	3604.970	4522.640	270.83	22554.92	41,456	58,854	2,781	304,000
United Kingdom	3277.012	2902.296	563.64	16173.30	41,845	29,020	11,498	137,195
France	2525.309	2319.842	316.25	9009.44	14,243	13,045	5,335	49,693
Rest of Europe	4615.975	6688.475	270.83	22554.92	56,573	93,604	2,781	304,000
Japan	2312.199	1781.85	594.3137	8702.235	6,754	3,871	1,061	15,919

TABLE 2.4. Descriptive statistics of total assets[a] and employment[b].

	Total assets				Employment			
	Mean	SD	Min	Max	Mean	SD	Min	Max
Rest of world	3277.624	2855.36	448.56	10213.00	35,982	26,961	9,000	85,000
Total	3843.414	5046.339	270.83	38528.00	37,720	49,757	1,061	304,000

Source: Authors' calculations based on AGRODATA.

[a]In million current US$ for each year.
[b]Total number of employees.

TABLE 2.5. Rates of growth of the world's 100 largest food and beverage multinationals, by home country: Assets, employment, and sales, 1985-1989 and 1990-1996.

	Growth of total assets				Growth of employment				Growth of sales			
	Mean	SD	Min	Max	Mean	SD	Min	Max	Mean	SD	Min	Max
1990-1996												
United States	7.14	13.44	−36.55	105.07	1.12	11.88	−92.13	30.12	5.58	10.38	−41.14	56.41
Europe	10.54	23.12	−30.51	170.82	5.66	27.72	−43.78	346.27	9.21	19.04	−42.17	126.50
United Kingdom	4.58	14.04	−30.51	45.25	0.47	12.85	−43.78	49.11	5.68	17.17	−42.17	99.91
France	14.36	19.92	−9.17	86.65	5.39	15.24	−26.46	55.22	9.86	13.09	−12.39	49.27
Rest of Europe	14.54	29.04	−22.39	170.82	10.63	38.75	−21.44	346.27	12.21	21.96	−21.21	126.50
Japan	8.14	12.32	−22.96	60.64	6.23	19.26	−3.89	81.33	7.40	12.10	−15.96	69.34
Rest of world	7.25	18.08	−27.75	64.82	0.50	7.27	−19.32	16.42	6.01	16.09	−42.85	60.61
Total	8.42	18.43	−36.55	170.82	3.69	21.71	−92.13	346.27	7.51	15.34	−42.85	126.50
1985–1989												
United States	17.13	25.52	−21.66	190.88	3.74	15.51	−58.93	61.53	12.57	18.77	−28.37	156.66
Europe	28.65	33.79	−11.02	253.31	5.94	17.57	−31.30	82.08	20.03	23.30	−29.00	171.60
United Kingdom	31.84	41.41	−10.87	253.31	5.19	19.84	−31.30	82.08	19.70	28.24	−29.00	171.60
France	36.31	26.87	4.78	114.12	9.36	21.43	−28.71	75.87	25.58	22.07	−7.84	86.47
Rest of Europe	19.86	18.85	−11.02	61.81	5.13	10.04	−7.42	38.43	17.59	14.39	−4.67	51.69
Japan	21.86	22.76	−11.62	79.27	1.41	11.10	−33.91	26.07	17.61	26.68	−67.86	135.80
Rest of world	15.14	10.56	−4.32	33.07	2.80	19.41	−27.72	55.64	24.93	41.98	−5.56	194.31
Total	22.67	28.78	−21.66	253.31	4.32	16.18	−58.93	82.08	17.41	24.58	−67.86	194.31

Source: Authors' calculations based on AGRODATA.

ECONOMETRIC SPECIFICATIONS

As mentioned, the aim of this chapter is not only to describe the pattern of geographical and industrial diversification but also to provide empirical evidence on the impact of diversification on firm growth. We use panel-data econometric techniques. This is the most efficient use of the data set, because it combines data for individual units (firms) with data for different time periods.

Our econometric model is built on two functions identified by Cantwell and Sanna-Randaccio (1993),[9] based on the work of Downie (1958) and Penrose (1959). We will use a reduced-form equation of the system. Since it is a standard finding that firm size and growth follow a lognormal distribution, the system can be specified as follows:

$$Log(GROWTH_i) = \alpha_0 + \alpha_1 Log(PROF_i) \qquad (2.3)$$

$$Log(PROF_i) = \beta_0 + \beta_1 Log(GROWTH_i) + \beta_2 Log(SIZE_i) \\ + \beta_3 X_i + \beta_4 Z_i + \beta_5 M_i \qquad (2.4)$$

Substituting Equation 2.4 into Equation 2.3, we obtain the reduced-form equation of the system, which has the following form:

$$Log(GROWTH_i) = \gamma_0 + \gamma_1 Log(SIZE_i) \\ + \gamma_2 X_i + \gamma_3 Z_i + \gamma_4 M_i \qquad (2.5)$$

Model Specification

After a careful examination of the data we decided to use a one-way error component model. The model is of the following form:

$$y_{it} = \alpha + X'_{it}\beta + Z'_i\gamma + M'_{it}\delta + u_{it} \qquad (2.6)$$

where $i = 1, \ldots, N$ denotes firm i and $t = 1985, \ldots, 1996$ denotes the year, and y_{it} is the dependent variable measuring growth[10]; α is a scalar; β is a 3x1 vector of coefficients of the control variables; X_{it} is the ith observation on the three control variables of size ($SIZE_{it}$), leverage (LEV_{it}), and capital intensity (CAP_{it}). Their description can be found in the following list.

$SIZE_{it}$ (firm's size): Logarithm of the total number of employees for firm i and time t

LEV_{it} (leverage ratio): Total debt/capital of the firm

CAP_{it} (capital intensity ratio): Fixed assets/total number of employees

Z_i is the measure of industrial diversification as represented by the variables in the following list.

$WITHINCORE_i$: Measures diversification within core business, i.e., dairy products, oils and fats, alcoholic drinks. It is the actual number of four-digit sectors in which the firm is active within the food and beverage industry. Indicates related diversification.

$AGRIC_i$: Actual number of agricultural sectors in which the firm is active (UN-SIC Codes: 1110, 1210, 1300, 1301, 1302) (agriculture, horticulture, animal husbandry, viticulture, pisciculture, aviculture, silviculture, fisheries, and production of seeds). Indicates vertical relatedness.

$RETAIL_i$: Actual number of retail sectors in which the firm is active (UN-SIC Codes: 6210, 6220, 6300, 6310) (retailing, supermarkets, hypermarkets, restaurants, and pubs). Indicates vertical relatedness.

$TECHN_i$: Actual number of technology-related sectors in which the firm operates independent affiliates (UN-SIC Codes: 311280, 832020, 832021, 832030, 9320, 9330) (technological services to other companies, biotechnology, veterinarian services to farms, production of microbiological products, and research centers with the status of independent affiliates). Indicates vertical relatedness.

$OTHER_i$: Number of nonfood industries and services in which the firm is active, excluding of course vertical-related industries. Indicates nonrelated diversification.

$FOODSA_{it}$: Food sales/total sales ratio of the company. Approximates the economic weight of related activities.

M_{it} is the measure of geographical diversification and is captured by the variables presented in the following list.

FDIV$_{it}$: Foreign affiliates/total number of affiliates. Gives a measure of the importance of geographical diversification within the multinational group.

FCOU$_{it}$: Number of foreign countries in which the firm is present. Country spread. Gives a measure of the geographical dispersion of the activities.

FAFC$_{it}$: Number of foreign affiliates/number of foreign countries in which the firm is active. It approximates the complexity of foreign networks in the company.

γ and δ are, respectively, the coefficients of the diversification measures. $u_{it} = \mu_i + \varepsilon_{it}$, where μ_i is the unobservable individual specific effect and ε_{it} is the remainder disturbance. Moreover μ_i is time invariant and accounts for any individual specific effect that is not included in the regression. The estimation method is displayed in the Appendix.

The study period, as shown, includes two subperiods: one of quick expansion, the other of steady expansion.[11] The time variable is obviously a major environmental determinant of growth on *all* FBMs over the study period. However, we do not aim to determine all the determinants of growth for this multinational sector. Rather, we try to understand the effect of diversification on the expansion of a company. We select control variables that could affect the relationship between diversification and growth at the company level, such as the debt burden of a firm. We do not include control variables that affect all companies in our sample similarly, such as time dummies.[12]

Results

To test the impact of each type of diversification on firm growth, we display the results of the econometric analysis in three different sections. Tables 2.6 and 2.7 present the results related to geographical and industrial diversification, respectively. Table 2.8 contains the results on the impact of both types of diversification on firm growth. In general, control variables, such as firm size, influence firm growth negatively, while the capital intensity has a positive impact (see "EQ1"

TABLE 2.6. Results on the impact of geographical diversification on firm growth.

	EQ1	EQ2	EQ3	EQ4	EQ5
Log(SIZE$_{it}$)	-0.18606[a]	-0.19701[a]	-0.22781[a]	-0.18666[a]	-0.22951[a]
	(0.03017)	(0.03069)	(0.02991)	(0.03023)	(0.03039)
LEV$_{it}$	-0.00297	-0.00261	-0.00303	-0.00297	-0.00286
	(0.00407)	(0.00407)	(0.00398)	(0.00407)	(0.0039)
CAP$_{it}$	1.32953[a]	1.31845[a]	1.21480[a]	1.33182[a]	1.2028[a]
	(0.11242)	(0.11237)	(0.11091)	(0.11263)	(0.11166)
FDIV$_{it}$		-0.24530[b]			-0.12936
		(0.14104)			(0.14214)
FCOU$_{it}$			0.00762[a]		0.0077[a]
			(0.00126)		(0.00131)
FAFC$_{it}$				0.00737	-0.01967
				(0.01822)	(0.01849)
C	0.20215	0.52994	0.50412[b]	0.19110	0.68036
	(0.30658)	(0.35747)	(0.30036)	(0.30791)	(0.35558)
R-square	0.4870	0.4914	0.5215	0.4883	0.5174
Wald chi²	207.48	211.64	258.49	207.37	257.85
σ$_\mu$	0.43304826	0.430611	0.408733	0.433382	0.41338
σ$_\varepsilon$	0.23158141	0.231008	0.224211	0.231717	0.22407

TABLE 2.6. Results on the impact of geographical diversification on firm growth.

	EQ1	EQ2	EQ3	EQ4	EQ5
Breusch-Pagan	1709.65	1659.61	1549.02	0.231717	1537.25
Prob > chi2	0.000	0.000	0.000	0.000	0.000
No. of observations	736	736	736	736	736

Note: Standard errors in parenthesis; dependent variable: GROWTH$_{it}$; method of estimation: unbalanced random effects estimator (corrected for heteroscedasticity).

[a] Significant at 1 percent.
[b] Significant at 10 percent.
[c]Refers to the probability of accepting the H_0 $\sigma_\mu = \varnothing$.

TABLE 2.7. Results on the impact of industrial diversification on firm growth.

	EQ1	EQ2	EQ3	EQ4	EQ5	EQ6
$Log(SIZE_{it})$	-0.18606[a]	-0.18599[a]	-0.19222[a]	-0.16966[a]	-0.19792[a]	-0.17034[a]
	(0.03017)	(0.03008)	(0.03064)	(0.03133)	(0.02937)	(0.03066)
LEV_{it}	-0.00297	-0.00281	-0.00295	-0.00294	-0.00289	-0.00273
	(0.00407)	(0.00405)	(0.00408)	(0.00406)	(0.00407)	(0.00404)
CAP_{it}	1.32953[a]	1.34276[a]	1.32748[a]	1.35772[a]	1.33427[a]	1.38127[a]
	(0.11242)	(0.11217)	(0.11213)	(0.11327)	(0.11022)	(0.11062)
$FOODSA_{it}$		0.23702[a]				0.22942[a]
		(0.09528)				(0.09417)
$WITHINCORE_i$			0.00363			-0.00373
			(0.01230)			(0.01229)
$AGRIC_i$					0.01799	0.04419
					(0.04926)	(0.05039)
$RETAIL_i$					-0.11069[a]	-0.08145[b]
					(0.03836)	(0.03969)
$TECHN_i$					0.25934[a]	0.32117[a]
					(0.06166)	(0.06777)
$OTHER_i$				-0.02468[c]		-0.04205[a]
				(0.01326)		(0.01401)
C	0.20215	0.0763	0.23781	0.14546	0.28578	0.00900

TABLE 2.7. Results on the impact of industrial diversification on firm growth.

	EQ1	EQ2	EQ3	EQ4	EQ5	EQ6
	(0.30658)	(0.30961)	(0.30591)	(0.30754)	(0.29524)	(0.30312)
R-square	0.4870	0.4905	0.4967	0.4438	0.5754	0.5463
Wald chi^2	207.48	214.94	211.35	211.13	257.36	272.67
Prob > chi^2	0.000	0.000	0.000	0.000	0.000	0.000
σ_μ	0.43304	0.43041	0.42344	0.43426	0.38862	0.39074
σ_ε	0.23158	0.22992	0.23158	0.23158	0.23158	0.22992
Breusch-Pagan	1709.65	1647.62	1526.41	1633.01	1253.28	1210.51
Prob > chi^{2d}	0.000	0.000	0.000	0.000	0.000	0.000
No. of observations	736	736	736	736	736	736

Note: Standard errors in parenthesis; dependent variable: GROWTH$_{it}$; method of estimation: Unbalanced random effects estimator (corrected for heteroscedasticity).

[a] Significant at 1 percent.
[b] Significant at 5 percent.
[c] Significant at 10 percent.
[d] Refers to the probability of accepting the H$_0$:σ_μ=∅.

TABLE 2.8. Results on the impact of geographical and industrial diversification on firm growth.

	EQ1	EQ2	EQ3
Log(SIZE$_{it}$)	–0.22883[a]	–0.2056[a]	–0.17637[a]
	(0.03004)	(0.03062)	(0.03116)
LEV$_{it}$	–0.00303	–0.00300	–0.00240
	(0.00399)	(0.00396)	(0.00403)
CAP$_{it}$	1.2122[a]	1.2634[a]	1.37640[a]
	(0.11897)	(0.11036)	(0.11095)
FDIV$_{it}$			–0.24034[c]
			(0.14482)
FCOU$_{it}$	0.00772[a]	0.00775[a]	
	(0.00141)	(0.00140)	
FAFC$_{it}$			–0.00761
			(0.01917)
FOODSA$_{it}$	–0.01680	–0.01866	0.22710[b]
	(0.10406)	(0.1026)	(0.09865)
WITHINCORE$_i$		–0.005928	–0.00139
		(0.011988)	(0.01252)
AGRIC$_i$		0.04729	0.04636
		(0.04909)	(0.05092)
RETAIL$_i$		–0.07206[c]	–0.07965[b]
		(0.03870)	(0.04003)
TECHN$_i$		0.3062[a]	0.31052[a]
		(0.0660)	(0.06864)
OTHER$_i$		–0.04774[a]	–0.04660[a]
		(0.01369)	(0.01445)
C	0.5216[c]	0.4111	0.30353
	(0.30940)	(0.30468)	(0.35918)
R–square	0.5231	0.5558	0.5444
Wald Chi^2	258.66	316.21	272.98
Prob > chi^2	0.000	0.000	0.000
σ_μ	0.40780	0.37816	0.39455

TABLE 2.8. Results on the impact of geographical and industrial diversification on firm growth.

	EQ1	EQ2	EQ3
σ_ε	0.22429	0.22429	0.22967
Breusch–Pagan	1528.04	1203.59	1185.00
Prob > chi[d]	0.000	0.000	0.000
No. of observations	736	736	736

Note: Standard errors in parenthesis; dependent variable: GROWTH$_{it}$; method of estimation: unbalanced random effects estimator (corrected for heteroscedasticity).

[a] Significant at 1 percent.
[b] Significant at 5 percent.
[c] Significant at 10 percent.
[d] Refers to the probability of accepting the H_0:σ_μ =∅.

column in Table 2.6). The leverage ratio is insignificant in all specifications.

Growth and Geographical Diversification

All the measures of geographical diversification except the average number of affiliates by country (FAFC$_{it}$), which captures the degree of integration within foreign countries, are statistically significant when used alone (Table 2.6). Country spread (FCOU$_{it}$) has a positive impact on growth, though the coefficient of FCOU$_{it}$ is very small. Operations in a great number of countries probably make the firm less vulnerable to country-specific factors and risks. Capturing the increased complexity of FBM networks, the ratio of foreign affiliates to total affiliates (FDIV$_{it}$) has a negative, though limited, impact on the rate of growth, given that the coefficient is small. Finally, as can be seen from column "EQ5" in Table 2.6, when we use the number of foreign countries as an explanatory variable, it absorbs all the significance of the other two measures of geographic diversification, while the sign of FDIV$_{it}$ remains negative, but the variable is insignificant.

Growth and Industrial Diversification

Table 2.7 reports the impact on growth of various types of industrial diversification.

The results support the theoretical framework we used before. Related diversification seems to positively affect the growth of a firm. The coefficient of FOODSA$_{it}$ is positive and statistically significant in all cases (some of the combinations are not displayed here). The positive sign of WITHINCORE$_i$ in column "EQ2" of Table 2.7 further supports this result, though it lacks statistical significance. Non-related diversification, on the other hand, measured by OTHER$_i$, increases network complexity and uncertainty, as the firm enters new markets where it has no experience or advantages. This likely leads to lower rates of growth. Furthermore, diversification into technologically related activities (TECHN$_i$) leads to increasing growth rates. This result could seem surprising. While the rationale of the international expansion of multinationals lies in the deployment of proprietary assets, the most important of which is knowledge (Caves, 1996), many highly profitable FBMs are hardly innovative (Christensen, Rama, and von Tunzelmann, 1996). Moreover, rates of R&D to sales are comparatively low, even among the largest companies in this industry (Grunert et al., 1995). Finally, other aspects of vertical integration in the firm's value chain, such as retail and agricultural operations, have no impact or even a negative impact on the growth of the firm.

Growth and Both Types of Diversification

Finally, we will assess the impact on growth of both geographic and industrial diversification. Table 2.8 presents the results for growth, controlling for both types of diversification. Our previous conclusions remain unchanged, which shows that our statistical results are robust. Yet the introduction of country spread (FCOU$_{it}$), much more important for inducing growth, makes the weight of core businesses (FOODSA$_{it}$) irrelevant. By contrast, FOODSA$_{it}$ counterbalances the negative influence on growth of the two other variables measuring geographic diversification, i.e., FDIV$_{it}$ and FAFC$_{it}$. The correlation coefficient for our measures of industrial or geographical diversification shows that these two measures of diversification are not correlated. Furthermore, we tested this model with country dummies, but again the results remained unchanged.

Though the independent variables explain around 50 percent of the variability of the rate of growth in FBMs and the Wald test shows that the regression equations are statistically significant, our analysis should be viewed as a study of the effects of diversification on growth when other determinants are controlled for, rather than as an attempt to model growth in FBMs.

We also tested the model with profitability as the dependent variable. Though firms tended to enjoy similar profit margins (5.0 to 4.6 percent) over 1985-1994, their profits on assets fell from 8.4 to 6.4 percent. Their profits on equity also dropped, from 16.3 to 12.7 percent. The results of the econometric analysis were poor. The cause could be our use of after-tax profit data, the only data available, since differences in tax systems are likely to distort results. Most international comparisons use gross profit data instead.

CONCLUDING REMARKS

With some exceptions, the world's 100 largest food and beverage multinationals grew more slowly (and became less profitable) by the late 1990s.

Over different phases of the business cycle, firms that grew faster tended to be relatively small and capital-intensive multinationals that had avoided diversification into retailing and into noncore businesses, i.e., nonfood products (after allowing for leverage). They may or may not have entered a large number of food and drink industries since related diversification practically did not affect growth. By contrast, they had spread into food-related technological activities, such as biotechnology, specialized services, and microbiological products. They tended to operate in a great number of countries, yet their share of foreign to total affiliates remained relatively low. When they did have a strong international projection and complex involvement at the host-country level, they rather opted for concentration in their core business, keeping a substantial share of food in their total sales.

FBMs coming from different home countries differ regarding diversification strategies, both geographic and industrial. To some extent, such strategies and national traits of the companies explain why some national groups perform faster than others. For instance Japanese multinationals, the fastest growers in the top group, are small, scarcely diversified into nonfood, and involved instead in technological activities related to agricultural and food production. However, country dummies are not good predictors of growth. In fact, correspondences should not be pushed too far. For instance, Japanese firms—as we said, quick growers—often diversify into agricultural activities, which are not especially conducive to fast growth.

Although we use different measures, our results coincide with those of Markides and Williamson (1994) and Fan and Lang (2000) in that we also reject the hypothesis that relatedness always facilitates good performance. Our results also show, as do Fan and Lang's, that some types of vertical relatedness are associated with poor performance. Our findings also support Frankho's (1989) point of view in that R&D activity is a very important predictor of corporate growth in food and beverage multinationals. Our research shows, as did Tallman and Li (1996), that country scope is positively related to performance (their control variables are similar to ours).

In spite of the unique traits of FBMs, many factors of expansion are quite similar in such firms and other multinationals. A relatively small size, reduced unrelated diversification, and substantial involvement in high-tech activities contribute to fast growth among FBMs and other multinationals (Cantwell and Sanna-Randaccio, 1993; Geringer, Beamish, and daCosta, 1989; Siddarthan and Lall, 1982). Our finding concerning high-tech activities in FBMs, surprising given the relatively low R&D/sales level in this industry, imply that diversification into such activities could also positively influence the expansion of multinationals in other traditional industries.

Researchers who analyzed other multinationals did not appear to agree on the effect on performance of geographical diversification. What seems to matter among FBMs is country spread. Over the study period, the FBM actually multiplied its geographic markets. At the same time, it checked the development of networks in host countries and maintained the same relative emphasis of foreign versus domestic activities. We interpret that firms won a great many markets in order to limit the expansion of their rivals without deepening their own involvement in each host country. This strategy was reflected in the specific route to expansion of FBMs, which is built on the presence of the firm in a multiplicity of foreign markets. Country spread influences it positively, and FBMs entering many geographic markets expand quickly. This result confirms that market proximity is crucial for this type of multinational because they need to interact closely with a variety of national consumers and are not likely to serve foreign markets through exports (Pearce, 1993). In addition, the small initial investment of some FBMs in "new" markets, such as China (Rama, 1992), suggests that the pioneer foreign company has often been able to deter entry with a limited presence, while acquiring knowledge of the new market and creating conditions for growth at the local level. By contrast, a substantial emphasis on foreign versus domestic activities (as measured by the foreign affiliates/total affiliates ratio) negatively influences the growth of the FBMs because such structure probably multiplies the cost of organization.

Related industrial diversification, as measured by number of related businesses, makes a modest contribution to growth both in other multinationals (Pearce, 1993) and in FBMs. However, a large F&B sales/total sales ratio is likely to induce development in the latter. In spite of the beneficial effects of concentration in core businesses, few FBMs intensified such a strategy from 1985-1989 to 1990-1996. Although sales of nonfood products, as a proportion of total sales in the top group, fell from 25.7 percent over 1981-1988 to only 10.2 percent over 1990-1996, changes chiefly took place during the early 1980s. Reducing their noncore involvement over the 1990s, Japanese FBMs are an exception. This strategy could be one of the keys to understanding their dynamism, even through low phases of the business cycle.

In other cases, our findings on FBMs could suggest future changes in other multinationals' diversification strategies. FBMs that diversified into agriculture or retailing were not especially fast performers over the study period, though vertical diversification had played its role in the past (Dunning, 1993; Rama, 1998). Our results imply that the internalized markets of multinationals could lose importance with time, as external markets for goods and services work better. Although diversification into agriculture and retailing are rarely important for other multinationals, our results suggest that other internalized markets for inputs and services could have only a transitory effect on the route to growth of companies.

Other inquiries remain for future research, such as exploring finely grained diversification, testing the curvilinear relationship between related innovation and growth (our model has tested only linear relationships), and studying the interaction of industrial and geographic diversification in FBMs. Qualitative aspects, not investigated here, also seem important.

APPENDIX

Characteristics of the Sample

In 75 percent of cases, the series of data are complete for the twelve years between 1985 and 1996. In another 16 percent, data for 1985-1996 are available but information for one to three intermediate years is missing. Thus 91 percent of the sample consists of FBMs that were in the top group over 1985-1996. By contrast, nine FBMs joined them between 1987 and 1990 (and were still in the top group in 1996). Their series of data cover a shorter span than those of the other ninety-one; the smallest is a seven-year series (one company). Newcomers are mostly companies from "new" source countries for FDI in this industry, such as Germany and Norway. Conversely, a few companies appearing in the top group by the mid-1980s were not included in our sample because, having dropped out *before* 1990, their series of data had fewer than seven years. Leaving the top group does not necessarily mean the death of the company. Some that are still in business were in the top group only temporarily by the mid-1980s. Most of the multinationals that left the top group, however, were U.S. multinationals acquired by financial groups or other companies in the 1980s. Finally, in a few cases, we were unable to find out why the enterprise dropped out, though we traced company names to make sure that the FBM was not active under a different name. No attempt has yet been made to investigate newcomers, incumbents, or companies that dropped out over diversification strategies.

Difference of Means Test

Table A2.1 uses a difference of means test to check whether the country spread of FBMs, classified by their home country, varied between 1985-1989 and 1990-1996.

Estimation Method

Random versus Fixed Effects

It is a commonly addressed question in panel data whether one should treat the individual effects (μ_i) as specific or random. In this study we

TABLE A2.1. Differences in means for number of foreign countries, by home country of the FBM.

	Number of foreign countries				
	1985-1989		1990-1996		
Home country	Mean	SD	Mean	SD	T-stat
United States	17	13	23	20	3.04[a]
Europe of which:	16	15	20	18	2.43[a]
United Kingdom	10	9	13	11	1.99[b]
France	18	11	24	11	2.23[a]
Rest of Europe	23	20	25	23	0.58
Japan	6	5	9	6	4.12[a]
Rest of world	15	11	16	13	0.37
Total	13	13	18	17	5.23[a]

[a]Significant at 1 percent.
[b]Significant at 5 percent.

will treat the effects as random for three main reasons. The first comes from the nature of the sample, which contains data on the world's 100 largest FBMs. Thus N (number of observations) is large enough to lead to an important loss of degrees of freedom. Moreover there is no clear explanation, since all the included firms share almost the same characteristics, why one should treat the effects as individual specific. Though this argument is quite convincing, it is not enough to support the use of random rather than fixed effects.

The second argument comes out of the nature of this study. The main purpose is to explore possible linkages between diversification and growth. What matters most is to estimate the model with some measures of diversification. We measure industrial diversification with two types of measure. One is time variant ($FOODSA_{it}$) and the other, due to AGRODATA limitations, is individual specific but time invariant. This is not a serious problem since the product portfolios of firms are usually stable over the long run (Dosi, Teece, and Winter, 1992). Moreover, since our time-variant variable, i.e., food sales/total sales, remained stable throughout the period, there are reasons to believe that the product portfolio also did. By attempting to estimate our model using fixed effects, we cannot use measures of industrial diversification because these will be collinear with the individual effects. On the other hand, estimating our model by random effects enables us to include the measures of industrial diversification. Finally, we are interested in these firms not in themselves individually but as random draws from a larger population of FBMs, even though the sample is censored because of being limited to the top 100. Therefore we decided to treat μ_i as random. In every case the reported Breusch-Pagan LM test (Breusch and Pagan, 1980) supports our decision. Our three measures of geographic diversification, the dependent variable, and the control variables are time variant.

Unbalanced Panel Estimation

Although we tried to use the most complete dataset, some observations are still missing. The variance-covariance matrix must take into account the fact that t (time periods) are not the same for each individual unit (firm). The solution to this problem is quite easy. We must compute θ_i, the weight that random effects uses to transform the dependent variable, and allow for different time periods for each each observational unit.

Heteroscedasticity

The standard error component model assumes that the regression disturbances are homoscedastic with the same variance across time and individuals. This is a quite restrictive assumption, especially in a panel such as the one examined in this study. We already have a certain type of heteroscedasticity since we allowed the group size to vary. To correct for heteroscedasticity due to different groups we also allowed the disturbance variance of the group-specific effect component μ_i to vary across groups. Therefore θ_i becomes the following:[13]

$$\theta_\iota = 1 - \frac{\sigma_\varepsilon}{\sqrt{T_i \sigma_{\mu i}^2 + \sigma_\varepsilon^2}} \tag{2.7}$$

NOTES

1. As Germany accounted for 15.3 percent of the world's largest multinationals and 25.6 percent of FDI outward stock in 1997 (Ietto-Gillies, 2002), the small number of German firms among top FBMs is noteworthy.

2. Global industries are those in which affiliates are able to establish an international division of labor within the multinational (Porter, 1986). In such industries, affiliates specialize in a small range of products or in parts used in further processing by affiliates located in other countries. In multidomestic industries, by contrast, firms are unable to organize such networks, and competition with other companies takes place in a variety of domestic markets. Semiconductors and automobiles are

examples of global industries while F&B is often cited as a multidomestic industry (Cantwell and Sanna-Randaccio, 1993; Porter, 1986).

3. The foreign sales/total sales ratio is not available for all the sample; thus we opted for the foreign affiliates/total affiliates ratio.

4. In general, the pattern in FBMs is quite similar, by home country, to that found by Ietto-Gillies (2002) in a sample of 664 large multinationals in manufacturing and mining analyzed in 1997.

5. The FBMs average ratio of foreign to total affiliates (54 percent) is similar, for 1990-1996, to that observed by Ietto-Gillies (2002) in multinationals from all sectors (53 percent) for 1997.

6. AGRODATA offers annual exchange rates of different currencies against the dollar, which were used to convert the financial data in current US$. Taking means over time is like taking the average exchange rate of each period.

7. Growth of variable V is calculated as:

$$GR = \left[\left(\frac{V_t}{V_{t-1}} \right) \times 100 \right] - 100$$

8. Among the three variables considered in Table 2.5, the rate of growth of employment is, in general, the least dynamic, a situation suggesting that FBMs are becoming more capital intensive. This interpretation would agree with that of Christensen, Rama, and von Tunzelmann (1996) who analyzed a similar sample of companies from 1977-1981 to 1986-1989.

9. However, both the variables considered and the research purposes here are different from those in the Cantwell and Sanna-Randaccio (1993) paper.

10. Growth of sales (GROWTH$_{it}$) was calculated by subtracting the logarithm of the value of sales for period t-1 and firm i from the equivalent for the period t and firm i. The result is the logarithm of proportional growth plus one.

11. Though this question goes beyond the scope of our chapter, one reason for fast growth in the 1980s could be that investors perceived the F&B industry to be a stable, countercyclical industry (Rama, 1992), less subject to turbulence than the, by then, emerging high-tech industries. In fact, the F&B industry of the United States and the European Union performed better than other industries over the 1980s. Good performance attracted many institutional investors, such as pension funds, and increased temporarily the resources available to large F&B firms (Christensen, Rama, and von Tunzelmann, 1996). Together with a demand shift toward high value-added foodstuffs and a wave of M&A (see Chapter 1), this situation could have transitorily stimulated fast growth in FBMs over the 1980s. According to (Rastoin et al., 1998), in the 1990s, the restructuring concluded, FBMs resumed their previous, slower rhythm of growth. In the 1990s, the average multinational grew more dynamically (12 percent over 1990-1994) (Ietto-Gillies, 2002) than the FBMs (only 7 percent in 1990-1996).

12. Other studies aiming at isolating the effect of diversification on performance, not at estimating determinants of growth, also follow a selective research strategy concerning control variables (Markides, 1995).

13. We need to estimate $\sigma_{ui}^2 = s_i^2 - s^2$ where s^2 is the residual variance of the consistent estimator σ_ε^2 of the least-squares dummy variable (LSDV) model and

$$s_i^2 = \frac{\sum_{t=1}^{T}(e_{it} - \bar{e}_i)}{T_i - 1}$$

where e_{it} are the residuals of the OLS estimators. Furthermore, to avoid negative values of σ_μ^2 we used a methodology proposed by Baltagi (1995) and replaced negative values with zeros.

REFERENCES

Baltagi B. (1995). *Econometric analysis of panel data.* New York: John Wiley and Sons.

Breusch T. and Pagan A. (1980). The LM test and its applications to model specifications in econometrics. *Review of Economic Studies,* 47: 239-254.

Cantwell J. (1989). *Technological innovation and multinational corporations.* Oxford and Cambridge: Basil Blackwell.

Cantwell J. and Hodson C. (1991). Global R&D and UK competitiveness. In M. Casson (Ed.), *Global research strategy and international competitiveness* (pp. 133-183). Oxford and Cambridge: Basil Blackwell.

Cantwell J. and Sanna-Randaccio S. (1993). Multinationality and firm growth. *Welwirtschaftliches Archiv,* 129(2): 275-299.

Caves R. E. (1996). *Multinational enterprise and economic analysis* (Second edition). New York: Cambridge University Press.

Chandler A. D. J. (1990). *Scale and scope: The dynamics of industrial capitalism.* Cambridge, MA: Harvard University Press.

Christensen J. L., Rama R., and von Tunzelmann N. (1996). Study on innovation in the European food products and beverages industry. Monograph number 145. The European Commission: EIMS/SPRINT Brussels.

Connor J. M. (1997). Economic overview and research issues: On the convergence of food systems. *Agribusiness,* 13(2): 253-259.

Dass P. (2000). Relationship of firm size, initial diversification, and internationalization with strategic change. *Journal of Business Research,* 48: 135-146.

Davis R. and Duhame I. M. (1995). Diversification, vertical integration, and industry analysis: New perspectives and measurement. *Strategic Management Journal,* 13: 511-524.

Ding J. Y. and Caswell J. A. (1995). Changes in diversification among very large food manufacturing firms in the 1980s. *Agribusiness,* 11(6): 553-563.

Dörrenbächer C. (2000). Measuring corporate internationalisation: A review of measurement concepts and their use. *Intereconomics,* 35(3): 119-126.

Dosi G., Teece D., and Winter S. G. (1992). Toward a theory of corporate coherence: Preliminary remarks. In G. Dosi, R. Giannetti, and P. A. Toninelli (Eds.), *Technology and enterprise in a historical perspective* (pp. 185-211). Oxford: Clarendon Press.

Downie J. (1958). *The competitive process.* London: G. Duckworth.

Dunning J. H. (1993). *The globalisation of business.* London and New York: Routledge.

Fan J. P. H. and Lang L. H. P. (2000). The measurement of relatedness: An application to corporate diversification. *The Journal of Business,* 73(4): 629-660.

Frankho L. G. (1989). Global corporate competition: Who's winning, who's losing, and the R&D factor as one reason why. *Strategic Management Journal,* 10: 449-474.

Geringer J. M., Beamish P., and daCosta R. (1989). Diversification strategy and internationalization: Implications for MNE performance. *Strategic Management Journal,* 10: 109-119.

Geringer J. M., Tallman S., and Olsern D. M. (2000). Product and international diversification among Japanese multinational firms. *Strategic Management Journal,* 21: 51-80.

Grunert K. G., Harmsen H., Meulenberg M., Kuiper E., Ottowitz, T., Declerck F., Traill B., and Göransson G. (1995). Innovation in the food sector: Between technology-push and demand-pull. *Structural Change in the European Food Industries (within the EU AAIR programme).* Discussion paper number 10 (December).

Hitt M. A., Hoskisson R. E., and Kim H. (1997). International diversification: Effects on innovation and firm performance in product-diversified firms. *Academy of Management Journal,* 40(4): 767-798.

Ietto-Gillies G. (2002). *Transnational corporations: Fragmentation amidst integration.* London and New York: Routledge.

Institut Agronomique Méditerranéen de Montpellier (IAMM) (1990). *Les 100 premiers groupes agro-alimentaires mondiaux.* Montpellier, France: IAMM.

Markides C. C. (1995). Diversification, restructuring, and economic performance. *Strategic Management Journal,* 16: 101-118.

Markides C. C. and Williamson P. J. (1994). Related diversification, core competences, and corporate performance. *Strategic Management Journal,* 15: 149-165.

McDonald J. R. S., Rayner A. J., and Bates J. M. (1989). Market power in the food industry: A note. *Journal of Agricultural Economics,* 40(1): 101-107.

Padilla M., Laval G. G., Allaya M.-C., and Allaya M. (1983). *Les cent premiers groupes agro-industriels mondiaux.* Montpellier, France: IAMM.

Palich L. E., Cardinal L. B., and Miller C. C. (2000). Curvilinearity in the diversification-performance linkage: An examination of over three decades of research. *Strategic Management Journal,* 21: 155-174.

Palich L. E., Carini G. R., and Seaman S. L. (2000). The impact of internationalization on the diversification-performance relationship: A replication and extension of prior research. *Journal of Business Research,* 48: 43-54.

Pearce R. D. (1993). *The growth and evolution of multinational enterprise: Patterns of geographical and industrial diversification.* Cambridge, UK: Edward Elgar.

Penrose E. (1959). *The theory of the growth of the firm.* Oxford: M.E. Sharpe, Inc.

Porter M. E. (1986). Competition in global industries: A conceptual framework. In M. E. Porter (Ed.), *Competion in global industries.* Boston, MA: Harvard Business Press.

Prahalad C. K. and Doz Y. L. (1987). *The multinational mission: Balancing local demands and global vision.* New York and London: The Free Press.

Rama R. (1992). *Investing in food.* Paris, France: OECD Development Centre Studies.

Rama R. (1998). Growth in food and drink multinationals, 1977-1994: An empirical investigation. *Journal of International Food and Agribusiness Marketing,* 10(1): 31-51.

Rastoin J. L., Ghersi G., Pérez R., and Tozanli S. (1998). *Structures, performances et stratégies des groupes agro-alimentaires multinationaux.* Montpellier, France: AGRODATA.

Rugman A. E. (1996). Risk reduction by international diversification. In A. E. Rugman (Ed.), *The theory of the multinational enterprises: The selected scientific papers of Alan M. Rugman,* Volume 1 (pp. 58-62). Cheltenham, UK, and Vermont: Edward Elgar.

Sambharya R. B. (2000). Research note: Assessing the construct validity of strategic and SIC-based measures of corporate diversification. *British Journal of Management,* 11: 163-173.

Siddarthan N. S. and Lall S. (1982). The recent growth of the largest US multinationals. *Oxford Bulletin of Economics and Statistics,* 44(1): 1-13.

Stopford J. M. and Dunning J. H. (1983). *Multinationals: Company performance and global trends.* London: McMillan Publishers.

Tallman S. and Lee J. (1996). Effects of international diversity and product diversity on the performance of multinational firms. *Academy of Management Journal,* 38(1): 179-196.

Williamson, O. E. (1975) *Markets and hierarchies: Analysis and antitrust implications.* New York: The Free Press.

Chapter 3

Innovation in Food and Beverage Multinationals

Oscar Alfranca
Ruth Rama
Nicholas von Tunzelmann

INTRODUCTION

Though food and beverages (F&B) processing is usually considered a low-tech industry,[1] profitability and growth seem to depend, to a large extent, on the food firm's ability to innovate continually (Connor, 1981). Advertisement and product differentiation, the other pillars of competition in F&B, also depend, in part, on good technology and design of packaging—thus the importance of innovation for F&B processing firms (F&B firms).

In this chapter, we analyze technical and design innovation in multinationals, the world's most important innovators in the F&B field, and review some of the research on this topic. Earlier attempts to study innovation in food and beverage multinationals (FBMs) were parts of cross-sectional studies on multinationals or focused only on specific facets of the innovative behavior of FBMs, such as the globalization of their R&D activities (for a previous review, see Wilkinson, 1998). This is the first time that the many dimensions of innovation in FBMs are integrated with an ample empirical base, encompassing FBMs of all sizes, subsectors, and nationalities.

We argue that a small nucleus of innovative FBMs, chiefly rooted in home laboratories, heads the world's production of food technology. Though FBMs integrate different types of innovation and shift their research interests over time, the *population* of innovators itself, we claim, remains stable. The innovative behavior of FBMs displays unique traits vis-à-vis other multinationals.

The first difficulty encountered by researchers who analyze the innovative activities of multinationals is to find common denominators for international comparisons among companies. Multinationals fund their R&D projects with a plurality of currencies and develop their innovative activities in a variety of countries with different legislation about intellectual property, which could make comparisons of patents granted in *different* countries less reliable. In addition, it is difficult to obtain a long-term series of R&D expenditure information at the company level, especially for non-U.S. firms. Here, we analyze panel data on FBM patents. One of the advantages of patent statistics is they provide a "unique long-term time series of inventive efforts on a worldwide basis" (Freeman, 1994, p. 476). Furthermore, patents reflect with some accuracy other manifestations of technological change, such as innovative activities and R&D expenditures at the firm level (Acs and Audretsch, 1989; Bound et al., 1984). The common objections to patent analysis,[2] on the other hand, lose their importance in relatively homogeneous samples of firms, such as those analyzed in this study.

We homogenize the information by counting the foreign patents granted to companies in *one* particular country, the United States, given that patenting in the United States seems to accurately reflect the world's stock of technology (Soete, 1987). It could be argued that this methodology underestimates the *level* of activities of non-U.S. companies, less likely to patent in the United States. Convincing in samples with smaller companies, in our view the objection is less valid here. As all have foreign direct investment (FDI) in the United States, the giant non-U.S. FBMs in our sample are likely to protect their property rights in such a major market, even if patenting in the United States is more expensive and time-consuming than patenting only in their home countries. Moreover, because of increases in globalization, it seems plausible that, if anything, our methodology overstates the relative *rate of growth* of patenting in the United States by non-U.S. companies.

In this chapter, we start by supplying information on the empirical evidence used in the research and by analyzing the recent evolution of patented innovation in FBMs. Then we study the influence of such companies in the world's production of technology and the international spread of their innovative projects. We identify innovators and their strategies within the multinational agrofood sector and, finally, provide conclusions.

PATENTED INNOVATION IN FOOD AND BEVERAGE MULTINATIONALS

The number of patents granted to FBMs climbed after the early 1980s in a context of changes in the composition of the world's top group and of increased participation of European and Japanese FBMs in innovation.

Before considering the question in detail, let us examine the bases for the empirical evidence provided in this chapter. The research is based on a large database with information on 16,414 patents granted in the United States over 1969-1994. The last year for which data consolidated at the company level are at present available is 1994. However, as innovation changes slowly in F&B (Galizzi and Venturini, 1996), the panorama depicted by the data is likely to reflect the current situation.

The sample considers a variety of technological fields such as food proper, biotechnology, tobacco, etc. Since we are interested in the world's 100 largest FBMs (hereafter, the top group), the sample includes patents registered for a selection of companies in this group and their affiliates over the period. The firms in our sample are multinational F&B processors that are active in a variety of industries included in United Nations–Standard Industrial Classification (UN-SIC) number 31, such as confectionery, dairy products, canned specialties, spirits, etc. While all are food or beverage processors, a number of them also have agribusinesses and other concerns (see the Appendix for more details on the data).

Figure 3.1 shows the evolution of the patents granted to the companies in our sample from 1969 to 1994. The total number of patents dropped after 1976 to rise again after 1980, until the mid-1990s. The recovery of innovative activity in the early 1980s preceded the recovery of profits among the world's largest FBMs, the rate of which fell to 12.8 percent in 1982-1985 (Christensen, Rama, and von Tunzelmann, 1996). The pace of innovation accelerated again when profits climbed quickly to 16.7 percent in the second half of the 1980s.

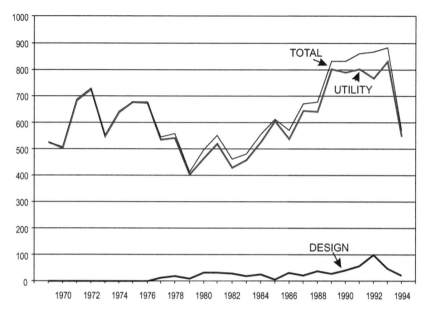

FIGURE 3.1. Number of patents granted to the largest F&B multinationals by type of patent, 1969-1994. (*Source:* Authors' calculations.)

U.S. FBMs are the most important patentors in the multinational agrofood sector, but their share tends to fall. In our sample, U.S. firms were those that innovated more, as a group, over 1969-1994, followed by European and Japanese companies (Table 3.1). The three sets of firms diverged, however, over this period since the number of patents awarded to Japanese and, to a lesser extent, European FBMs climbed, while those granted to U.S. firms declined. Among European FBMs, the group of British companies innovated less than at the beginning of the period, while the French, Dutch, Italian, and Swiss sets of companies patented more inventions in the United States than before, though often starting from modest points of departure. Part of the explanation for the dynamic increases of Japanese and European FBMs' patents in other parts of the world could lie in increasing tastes for Asian and European specialties. Meeting the U.S. demand for Japanese specialty foods, for instance, has been the most important strategy of Japanese FBMs in California (Jussaume and Kenney, 1993).

TABLE 3.1. Number, percentage, and growth of patents* granted in the United States to the world's largest FBMs.

Home country	No. of firms	No. of patents 1969-1994	Share (%) 1969-1974	Share (%) 1975-1989	Share (%) 1990-1994	Share (%) 1969-1974 to 1990-1994	Growth (%) 1969-1974 to 1990-1994
All countries	103	16,402	100.00	100.00	100.00	100.00	9.87
Argentina	1	38	0.00	0.31	0.23	0.28	–
Australia	2	8	0.03	0.06	0.05	0.05	100.00
Canada	6	181	2.34	0.86	1.10	0.53	–75.29
Japan	20	1,655	6.13	10.59	10.09	12.59	125.56
South Africa	1	85	0.00	0.50	0.52	1.03	–
United States	31	9,993	68.02	60.76	60.93	54.83	–11.44
Europe	42	4,442	23.48	26.92	27.08	30.71	43.68
Denmark	2	8	0.00	0.02	0.05	0.15	–
France	8	97	0.74	0.42	0.59	0.83	22.22
Italy	3	82	0.11	0.55	0.50	0.75	650.00
Sweden	2	42	0.44	0.25	0.26	0.10	–75.00
Switzerland	3	939	1.87	5.42	5.72	9.91	482.35
Netherlands	5	2,303	11.88	14.68	14.04	14.61	35.19
United Kingdom	19	971	8.44	5.59	5.92	4.35	–43.32

Source: Authors' calculations.
*Includes utility and design patents.

THE GLOBALIZATION
OF INNOVATIVE ACTIVITIES

In this section, we analyze two major aspects of technoglobalism and confirm their applicability to FBMs (for a survey, see Granstrand, Hakansson, and Sjölander, 1993). The term *technoglobalism* means, first, that multinationals now exert a substantial influence on innovation in specific industries worldwide. Second, the term also indicates that multinationals currently perform a large part of their innovative activities abroad. Before studying both aspects in FBMs, we briefly comment on their practical importance.

It has been claimed that trends toward technoglobalism could eventually make government measures and stimuli superfluous to technological development, given the tremendous impact of external agents, such as multinationals. Researchers opine that identifying where the multinational locates its R&D activities is relevant because host countries, for instance, could benefit from the local R&D activities of such firms (for surveys of the effect of multinationals on the innovative activities of host countries, see Cantwell and Janne, 1999; Dunning, 1994) or, alternatively, resent their presence. Home countries could, in turn, worry about losing the R&D facilities of their largest and more innovative companies to foreign nations. Are these worries justified? Based on empirical research, some authors maintain that the widespread globalization of technological production as a result of multinational relocation of R&D activities has been exaggerated because most of it still remains in the home country (Patel, 1995; Patel and Pavitt, 1991). As will be seen in this chapter, some of these worries could be justified in the case of FBMs.

To anticipate our results, we find that both aspects of technoglobalism apply to FBMs. First, similar to other multinationals in their respective industrial sectors, FBMs exert a large influence on the world's production of food and drink technology. Second, the FBM produces a much larger share of innovation in foreign locations than other multinationals; in this respect, the FBM's R&D investment is more internationalized. It is true that, as for other multinationals, FBMs produce most of their inventions in their home countries. There are examples of European national groups of FBMs, however, which have relocated most of their R&D to foreign nations.

FBMs Hold a Large Share of the World's Food and Drink Innovations

As stated, FBMs supply a large share of the world's innovation in food and drinks. Their contribution to sectoral innovation is around 50 percent, similar to that of the average multinationals in the world's production of innovations (Patel and Pavitt, 1991). According to Patel and Pavitt (1991), a set of seventy-five FBMs patented, in 1981-1986, 48.9 percent of the world's innovations produced in food and tobacco.[3] According to other estimates, the top group held around 51.3 percent of the world's patented innovations in F&B (tobacco included) over 1977-1994 (Alfranca, Rama, and von Tunzelmann, 2002). Single FBMs could dominate the production of innovation in specific countries or subsectors. For instance, Unilever (NL-UK) alone held, over 1969-1988, half of the oil and fat patents granted in Spain, a major producer of olive oil (Rama, 1991). On the other hand, FBMs control a greater share of the world's production of technology than of the world's production of food and drink, probably because most manufacturers in this industry are small enterprises with limited patenting activity. The top group of FBMs accounts for about 38 percent of the value of the world's production of processed foods and beverages (Rastoin et al., 1998), a proportion well below their 50 percent in the world's patented innovations.

Such companies supply a large share of the world's innovation in F&B and direct the innovation process in this industry (Organisation for Economic Co-operation and Development [OECD], 1979), but their participation is diminishing. We will turn to this next.

FBMs Innovate at a Slower Pace than Other Innovators in Food and Food-Related Fields

The share of FBMs in food and drink innovation is falling because other agents innovate more quickly. In this regard FBMs differ from other multinationals. In his review, Zanfei (2000) reports on research suggesting, for instance, that foreign direct investment is an "increasing component of overall scientific and technical activities" (p. 518).

One-nation firms, smaller food multinationals, universities, and research centers are innovating in the food and food-related fields at a faster pace than the top FBMs. Between 1969-1974 and 1990-1994, the number of food patents granted in the United States to the top FBMs grew 9.9 percent, while those awarded to all types of innovators in this field rose 13.3 percent.[4] In selected fields, differences in the pace of growth are enormous. While the number of biotech patents granted to the FBMs in our sample increased by 46.4 percent, those awarded to all kinds of patentors increased 649.0 percent.[5] By contrast, FBMs' sales increase more quickly than total world food sales. With such firms positioning themselves in dynamic segments of the international F&B industry and carrying out aggressive strategies of external growth, the global production of the top group increases, in value terms, at an average annual rate 3 to 4 percent higher than that of the world's production of processed foodstuffs and drinks (Rastoin and Tozanli, 1998).

Why, unlike other multinationals, do FBMs tend to control a diminishing share of the world's production of sectoral innovations? One reason could be that some FBMs still compete on the basis of advantages other than innovation. As shown by Rugman's (1987) study of a group of Canadian spirit and beer multinationals, some FBMs still rely more on country-specific resources and good distribution systems than on technological strengths. On the other hand, beyond the amount of innovations per se, an ownership (O) advantage of FBMs might be their capacity to combine bundles of different types of technology and control knowledge in upstream industries. These characteristics concerning the form and direction of innovation could be considered as part of Dunning's (1993) dynamic "add-on" strategy-related variables. We will come back to this strategic aspect of innovation in FBMs. Another reason for the faster growth of innovation in nonmultinational innovators versus FBMs could be that the latter are currently outsourcing some of their requirements, especially in biotechnological products. We will come back to this too.

A Substantial Internationalization of R&D

As stated, the fact that multinationals now perform part of their innovative activities abroad is another manifestation of technoglobalism. The share of patenting attributable to R&D activities in foreign locations is much higher in FBMs than in other multinationals. In a sample of multinationals pertaining to several industries, including F&B, the share of patents attributable to foreign locations oscillated around 10 percent of the total between 1969-1986 (Cantwell and Hodson, 1991). Patel (1995) finds a similar share for the subsequent 1985-1990 period. Both studies show that firms in different sectors display different propensities to develop foreign R&D activities and that, in this sense, FBMs are an outstanding example of techno-globalism. Cantwell and Hodson (1991) notice that the percentage of U.S. patents of the world's largest food firms attributable to research in foreign locations amounted to 24.0 percent by the mid-1980s. Patel (1995) observes that, by the beginning of the 1990s, food multinationals patented abroad 26.3 percent of their total innovations, and drink and tobacco multinationals 30.7 percent. Our results are comparable to those of the previously mentioned authors, though we consider a much larger span of time than they do and focus on North American and European Union FBMs only. In a subsample of fifty-four FBMs of such origins, which were granted more than 10,450 patents over 1969-1994, the companies perform 29.4 percent of their innovative activities abroad (see Table 3.2).[6]

Of the patents FBMs produce abroad, around 96 percent are utility patents and the remaining 4 percent are design patents, a type of innovation not investigated in cross-sectional studies but important for food firms owing to the impact of packaging in the marketing of foodstuffs and drinks. This distribution by types of patents is similar to that at home. As shown in the later discussion of technical and design innovation, both types of innovations are functionally associated within the FBM. They seem to be geographically associated too.

FBMs have tended to increase the percentage of their foreign versus their domestic R&D activities, while the average multinational has maintained almost the same proportion over 1969-1995 (Anastassopoulos et al., 1997; Cantwell and Janne, 2000). In spite of the growing importance of foreign innovative activities for FBMs as a group, 70 percent of patents still relate to domestic R&D and most FBMs research *only* at home. By the beginning of the 1990s, only 40 percent of the most important FBMs patented abroad, a much smaller share of multinationals than in other sectors (Anastassopoulos et al., 1997).

In host countries, FBMs' foreign subsidiaries are more influential than foreign subsidiaries from most other multinationals. The former accounted for nearly 22 percent of the innovations generated in their host countries in 1995-1999 (United States not included as an inven-

TABLE 3.2. Patenting abroad of the world's largest multinationals, by nationality of parent (%) (food and beverage multinationals, and multinational enterprises from all sectors).

| Home country | FBMs[a] | | | MNEs[b] |
	Utility	Design	Total	Utility
Total[c]	29.7	23.5	29.3	10.8
Canada	48.3	100.0	48.8	40.0
United States	9.7	13.5	10.0	6.92
Europe[d]	83.4	84.5	83.4	28.2
Denmark	0.0	50.0	12.5	N/A
France	11.7	50.0	12.9	12.6
Italy	39.0	14.3	27.3	14.0
Sweden	0.0	–	0.0	27.6
Netherlands	86.8	87.1	86.8	51.5
United Kingdom	84.3	92.6	84.8	46.3

[a]Authors' calculations. Food and beverage multinationals based in North America and the EU, 1969-1994.
[b]Calculated from Cantwell and Janne (2000). Multinational enterprises from all sectors and countries, 1969-1995.
[c]For FBMs, includes North America and the EU. For MNEs, includes all countries.
[d]For FBMs, includes EU countries. For MNEs, includes the EU, Norway, and Switzerland.

tor country) (Khanna and Singh, 2002). The average multinational in any industry, by contrast, accounted for 15 percent of the patenting activity in its host countries during the past thirty years. Even in developed host countries, foreign F&B affiliates are likely to rank at the top. In Italy, for instance, the most important patentors over 1967-1990 were the affiliates of seven foreign FMBs (Fanfani, Lanini, and Torroni, 1996). Unilever's subsidiaries alone were the leading patentors in seven out of seventeen Italian F&B industries.

Researchers often attribute the larger share of foreign innovative activities in FBMs to food firms (unlike other industrialists) needing to adapt their products to local tastes. This is an important reason behind the substantial internationalization of FBMs' innovative activities since, in spite of trends toward the homogenization of diets (Connor, 1997)—at least among Western countries—differences in local tastes are noticeable and persistent. Different national regulations concerning the safety of foodstuffs also induce FBMs, in their foreign laboratories, to adapt some products initially developed in the home country.

However, FBMs also internationalize their R&D investment for reasons other than adapting their products to local conditions. In addition, recent research suggests that their foreign laboratories could enjoy more autonomy than previously thought, an organizational aspect often related in the literature to the capacity to produce *new* products abroad. For instance, foreign laboratories active in the British F&B industry perceive the development, production, and marketing for the United Kingdom and/or wider markets of *new* products, *in addition* to the multinational group's existing range, as the most important role of the establishment (Pearce, 1999). By contrast, they see the adaptation of the group's products to British tastes as a secondary mission. This study suggests that, at least in some locations, adaptation functions could be less important than expected in the FBMs' foreign laboratories, while more creative and autonomous tasks become a priority. Another study suggests that the purely adaptive strategy could be only a transitional stage in FBM evolution (Anastassopoulos et al., 1997). The authors point to FBMs relying, even more heavily than multinationals in some high-tech industries, on "locally integrated laboratories," i.e., R&D organizations beyond the mere "support laboratory" devoted to adaptive development. Locally integrated laboratories, which instead promote applied research and innovation development, not only associate with subsidiaries that manufacture the same foodstuff range for local markets as the group but also link with mandated affiliates that operate as international (or pan-European) centers for producing specific products, often new, and distributing them across many countries. The current reorganization of foreign innovative activities could relate, therefore, to FBMs implementing new productive and marketing strategies at the supranational or the global level, beyond their traditional multidomestic pattern (see Chapter 2).

The idea of a historical technological evolution in FBMs' affiliates coincides, to some extent, with Reddy's (2000) theory of four waves in the globalization of R&D. The first would involve a few units meant to transfer technology from parents to affiliates. The second would focus on the adaptation of products to local markets. As stated, this has been one of the most important aims for the globalization of R&D in FBMs. The third wave would involve the worldwide acquisition of technology and, aided by the convergence of tastes, the development of products for major markets. Finally, the fourth would aim at recruiting scientists worldwide, even in developing countries, to reduce R&D costs. As will be seen, FBMs also seem to follow the last two steps of this scheme.

Abroad, FBMs also develop new interdisciplinary knowledge and basic research (Anastassopoulos et al., 1997), often in specialized R&D subsidiaries not attached to their foreign manufacturing centers (Rama, 1996b). In 1988, the world's top group already owned around eighty-four affiliates of this sort. Their independence and UN-SIC classes suggest they perform the "home-base augmenting" type of R&D, consisting basically of the absorption of local technological spillovers, a kind of foreign innovative activity already observed in other multinationals (Criscuolo, Narula, and Verspagen, 2002; Kuemmerle, 1999; Zanfei, 2000). FBMs holding independent laboratories devoted to generic knowledge in the fields of agriculture, husbandry, and food science are chiefly Japanese and, to a much lesser extent, British (Rama, 1996b). U.S. FBMs, though highly creative, avoid this organization of R&D activities. As shown in Chapter 2, diversification of the FBM into high-tech activities, as measured by the ownership of such independent laboratories, is a noteworthy factor in the global expansion of sales.

By the end of the 1970s, several studies already pointed to the international sourcing of technology by European FBMs (OECD,

1979). Japanese FBMs display similar trends. It was suggested that they invest in some U.S. businesses, such as wineries or sausage companies, chiefly to acquire technological and operational expertise that could be transferred to the parent, since such products are now becoming popular in Japan (Jussaume and Kenney, 1993). As for multinationals in chemicals, pharmaceuticals, mining, and materials, the FBM locates its foreign R&D activities chiefly in countries that are strong in its own fields of strength (Patel and Vega, 1999). This also suggests that the foreign laboratories of FBMs aim at absorbing foreign knowledge, as well as at adapting their foodstuffs to local tastes, as traditionally argued.

FBMs undertake innovative activities going beyond mere adaptation of foodstuffs to local markets, even in Southern Europe (Anastassopoulos et al., 1997) or in developing countries, such as Brazil, Hong Kong, India, and Singapore (Rama, 1996b; Reddy, 1993, 2000). Though most of the patenting abroad in our sample takes place in developed countries, we also found some innovative activity in affiliates located in Argentina and South Africa. This could coincide with Reddy's (2000) fourth wave.

Geographic decentralization of R&D could obey a multiplicity of strategic choices (Dunning, 1993) and is not necessarily associated with higher technological performance at the company level. Among a group of innovative FBMs within our sample of North American and EU firms, Allied-Lyons (United Kingdom), Hillsdown (United Kingdom), and Unilever (the Netherlands-United Kingdom), for instance, favor significant geographical decentralization of R&D, while Barilla (Italy), Pepsi (United States), and Kellogg's (United States) opt for centralized R&D structures and locate most of their innovative activities at home. All are successful companies ranking among the world's 100 largest FBMs, whatever their R&D organization.

Is the Globalization of Technology a European Phenomenon?

U.S. multinationals tend to concentrate their R&D activities at home more than other multinationals. When we focus on FBMs, U.S. companies also tend to research at home to a larger extent than FBMs based in other countries.

Moreover, it has been argued that the so-called globalization of R&D is actually a European phenomenon (Archibugi and Michie, 1995). First, European multinationals would be more prone than other multinationals to innovate abroad. Second, when researching abroad, European multinationals would prefer to locate their labs in other European countries rather than in the United States, Japan, or elsewhere. For these reasons, the internationalization of their R&D is sometimes interpreted as a regionalization rather than as a globalization process. Is this true of FBMs? The answer to this inquiry is yes and no. European FBMs are actually more prone than FBMs from other nationalities to internationalize their innovative activities, but they develop most of their foreign R&D in the United States, not in Europe.

As stated, U.S. FBMs locate only a small share of their R&D activities abroad. Though Europe is becoming a more important location for R&D activities (Cantwell and Janne, 2000), the twenty-eight American FBMs in our sample still source 90 percent of their research activities at home, a similar percentage to that in all types of U.S. multinationals (National Science Foundation, 1996). By contrast, the twenty-one EU companies in our sample source only 17 percent of their R&D activities in their respective home countries (see Table 3.2), a result comparable with other authors' findings. In a sample comprising FBMs based both in the EU and European Fair Trade Association (EFTA) countries, Cantwell and Janne (2000) observe that the companies in 1991-1995 carried out only 22.4 percent of their research activity at home.

Both U.S. FBMs and other U.S. multinationals research abroad to a lesser extent than their European counterparts. The comparison of our data on FBMs and those of Cantwell and Janne (2000)[7] on all types of multinationals show that companies pertaining to the same home country follow similar trends of R&D internationalization (see Table 3.2).[8] Among U.S. FBMs, the ratio of foreign to total patenting is larger than among other U.S. multinationals. However, when we compare such companies, the gap is not very large. By contrast, when we compare European FBMs and the average European multinational, the gap is extreme. European FBMs locate abroad 83 percent of their patenting, while the average European multinational locates only 28 percent. The parallel evolution of patenting abroad in FBMs and other multinationals from the same home country suggests that the internationalization of innovative activities could depend on the combination of a variety of factors at the company, sector, and national levels. Individual strategy, productive internationalization of the company, industrial characteristics (for instance, the need to adapt products in F&B), and those of the national system of innovation in the home country could each play a role.

Against all expectations, the foreign R&D activities of European FBMs are not attributable to increased regional integration, since only 22.2 percent are performed within Europe (Cantwell and Janne, 2000). Most of the rest are located in the United States. According to the study, F&B (including tobacco) is the sector in which the share of foreign innovative activities located by European multinationals in the United States is the largest (77.2 percent versus 53.1 percent in all industries). Understanding why European FBMs undertake such a large share of their innovative activities in the United States would require in-depth analysis of the nature of the innovative activities that FBMs develop in different sites, which is not attempted here.

Among European FBMs, differences in the preference for at-home R&D activities could be attributable to the company's position in the food chain and the technological base of the home country. For instance, Dutch F&B companies, such as Heineken or Melkunie (both in our sample), whose foreign production is supplementary to their exports, prefer to locate their labs in the Netherlands where they often cooperate with public research centers (Bijman, van Tulder, and van Vliet, 1997). Other major Dutch F&B companies are more likely to research abroad and choose, by contrast, "in-house" R&D. This seems to confirm that R&D abroad decreases the more the company relies on exports (as opposed to FDI) (Zejan, 1990).

Given that FBMs are able to absorb knowledge in a variety of geographic locations, the role of the technological base in the home country could be controversial. We turn to this question next.

The Influence of the Home Country

Do national systems of innovation in the home country influence the technological development of its multinationals? Since such companies currently internationalize their technological production, one argument goes, the role of their own home country as a source of knowledge must be small. Some authors demonstrate, however, that home countries could matter. Tracing patent citations, a study by Criscuolo, Narula, and Verspagen (2002) reveals that foreign laboratories of multinationals use both host-region and home-region knowledge. Alternatively, a study by Cantwell (1989), which deals with both trade and offshore production of firms from six industrial countries, shows that patterns of patenting in large companies are related to the technological specialization of the home country in early stages of internationalization but not later. Among FBMs, the home country could also exert an influence in the early stages of the internationalization of companies.

The performance of emerging FBMs in global markets is associated with the amount of effort devoted by the home country to innovation in a body of knowledge that includes food and food-related innovation (packaging, instruments, biotechnology meant for the agrofood chain, etc.), as measured by patenting activity in such fields by large and small firms, universities, and research centers (Rama, 1999). With evidence from a sample of 4,572 foreign patents granted to nationals from major OECD countries over 1969-1988 and economic data on ninety-six major FBMs, the study shows that, among smaller FBMs and newcomers, the most profitable tend to emerge from countries where the food chain is innovative. By contrast, among large, experienced FBMs, such an association could not be statistically confirmed. Direct links with the national innovation system seem to be diluted when the FBM acquires experience in the international market and greater size. In other words, within the world's top group, a highly global nucleus, influenced by a variety of geographic sources of technology, could coexist with FBMs relying largely on localized processes of innovation.

Bijman, van Tulder, and van Vliet (1997) point to a possible reason why technological links with the home country could be diluted over time. The internationalization of F&B companies, they contend, affects their R&D orientation. This could lead them to changing their demand for science and technology from national research centers. If such institutions want to continue servicing the FBM, the authors argue, they may have to change the scope of their activities. Their observations suggest that, if national innovation systems are unable to accomplish such changes, the FBM could become technologically footloose. Another reason could be that, over time, FBMs, like multinationals in other sectors, tend toward homogeneity (Hu, 1992; for arguments against homogeneity or isomorphism, see Hakanson and Nobel, 1993).

In this section, we have analyzed where the FBMs innovate. In the next, we identify innovators within the multinational agrofood sector and study the content of FBM research and the forces behind their innovative processes.

INNOVATORS AND THEIR STRATEGIES

FBMs combine different types of innovation with research in a variety of technological fields and change their R&D priorities through time. Yet the *population* of innovators itself, a small group of FBMs that innovate continuously and compete technologically with one another, tends to remain stable. We start by studying the agents of the innovative process.

A Handful of Innovators with a Visible Influence

Though several studies have analyzed the technological production of multinationals, few have identified innovators within each multinational industry, so we ignore whether other multinationals are homogeneous in terms of the production of innovation. FBMs are not. This subsection, which studies the production of innovation *within* the multinational agrofood sector, shows that not all FBMs are innovative.

Over 1977-1994, around half the FBMs in the top group were granted fewer than two utility patents and fewer than one design patent (Alfranca, Rama, and von Tunzelmann, 2001).[9] On the other hand, most FBMs remain continuously innovative only for short periods of time. According to a longitudinal analysis of patterns of innovation in FBMs over the same period, a relatively small core of persistent innovators and a large fringe of occasional inventors direct technological change in the multinational agrofood sector (Alfranca, Rama, and von Tunzelmann, 2004). Persistent innovators, who invent continuously for long periods of nearly two decades, account for a modest share of the top group (22 percent of firms) but supply a disproportionately high share of the total number of patents (80 percent). FBMs remaining innovative in the technical field also tend to remain innovative in design for long periods of time, probably an effective strategy for deterring entry and mobility in the sector given that *continually* launching new (Connor, 1981) and attractively packed food-stuffs is crucial for F&B companies. Deterring entry and mobility could be among the FBMs' objectives behind patent races in the multinational agrofood sector. Some of the companies more persistently innovative over the study period include CPC International, Coca-Cola, ConAgra, Itoham Foods, Mars, Nestlé, Sara Lee, and Unilever. In short, the multinational agrofood sector is highly heterogeneous as regards the agents of innovation. As shown in the following paragraphs, it is also heterogeneous concerning the technological fields that attract them.

R&D Priorities

FBMs undertake R&D in a variety of technological fields, not only in the food field (see Table 3.2). The following results refer to the top group (not to our subsample of North American and EU FBMs only). The pattern of technological diversification of FBMs depends on their pattern of product diversification and, to a much greater extent, on their need to control upstream techniques. FBMs diversify into non-food technology to a much greater extent than they diversify into nonfood products. Food and agriculture patents amount to around one-third of their innovations (see Table 3.3), which contrasts with around 80 percent of their sales value consisting of agricultural products, inputs for agriculture, food proper, and retailing (see Chapter 2). In addition, FBMs also innovate in technological fields disconnected from food (see "Other" row in Table 3.3, including textiles, vehicles, etc.), because many such companies are, as explained in Chapter 2, conglomerates highly diversified into nonfood industries.

This is not the only reason, nor the most important, why they diversify into nonfood technological fields. Another reason is that some FBMs, such as Ajinomoto and Ferruzzi (both in our sample), process by-products—for instance, glue, starch, or organic acids—from agricultural goods (GEST, 1986; Gonard et al., 1991). In part, this could explain FBMs patenting in technical fields that might seem, at first sight, unrelated to the food chain.

The most important reason why FBMs devote most of their efforts to food-related technology (machinery, instrumentation, bioengineer-

TABLE 3.3. Share of patents* granted to the world's largest FBMs by technological field, number of patents, and percentages (1969-1974 and 1990-1994).

Tech field	1969-1974	%	1975-1989	%	1990-1994	%	1969-1994	%
Total	3,638	100.0	8,769	100.0	4,007	100.0	16,414	100.0
Agriculture	36	1.0	117	1.3	47	1.2	200	1.2
Bioengineering	151	4.2	444	5.1	221	5.5	816	5.0
Chemistry	1,057	29.1	2,162	24.7	771	19.2	3,990	24.3
Drugs	113	3.1	787	9.0	490	12.2	1,390	8.5
Food	1,167	32.1	2,753	31.4	1,201	30.0	5,121	31.2
Instruments	114	3.1	363	4.1	256	6.4	733	4.5
Machinery	246	6.8	556	6.3	279	7.0	1,081	6.6
Other	754	20.7	1,587	18.1	742	18.5	3,083	18.8

Source: Authors' calculations.
*Includes both utility and design patents.

133

ing, etc.) is their aim to control the supply of the innovations needed to produce foods and drinks. In other words, FBMs internalize some markets for technology. The F&B industry depends, to a great extent, on upstream industries for innovation (Pavitt, 1984; Rama, 1996a). Unlike most food companies, however, the FBM produces part of such technology "intramurally." This characteristic of FBMs throws new light on the perception of the F&B industry as a supplier-dominated industry. In the very large food firms that dominate sectoral innovation, "external" technology is not purchased from suppliers, or at least not entirely. More important, the pattern of technological fields in FBMs confirms that the F&B industry, or at least its largest companies, are currently playing a much more active role in the integration of a variety of sciences and techniques than the "supplier-dominated industry" schema would suggest (Christensen, Rama, and von Tunzelmann, 1996).

The most important example of food-related fields is chemistry. In specific FBMs, the share of chemistry patents in total patents could be surprisingly high. For instance, in a sample of fourteen large food and eight drink and tobacco firms Patel and Pavitt (1997) find that chemical patents account, respectively, for 71 percent and 41 percent of the totals over 1981-1990. Accordingly, these authors classify chemistry, food and tobacco, and drugs and biotechnology as the three *core* technical fields in FBMs, while another fourteen technological fields are considered less important to these firms. Though in larger samples of FBMs, such as is analyzed here, the share of chemistry patents is smaller (see Table 3.3), it remains substantial. However, the types of upstream technologies in which FBMs are interested are changing according to the market and to new ways of solving technological problems in this industry (Christensen, Rama, and von Tunzelmann, 1996). FBMs tend to produce a smaller proportion of chemical patents than they did previously; this tendency is in accordance with the generalized substitution of biotechnological and other techniques that are more acceptable than chemical techniques in food production for some consumers. At the same time, they have become more interested in other technological fields. In a sample of 106 large FBMs, the combined share of patents in four fields related to new technologies—biotechnology, drugs, instruments, and electronics—rose from 13.3 percent to 20.6 percent from 1969-1974 to 1990-1994 (von Tunzelmann, 1998). FBMs are increasing their research efforts in the pharmaceutical field especially, because consumers demand more sports drinks or new healthy foodstuffs containing minerals, vitamins, and other additives. Companies such as Unilever and ADM (both in our sample) provide other food processors with hundreds of flavors, sweeteners, and artificial vitamins (Nicolas, 1996), often obtained thanks to biotechnology.

All multinationals tend to innovate in a variety of fields, well beyond their product range, in order to take care of technological interdependencies and opportunities in new techniques (Patel and Pavitt, 1997). FBMs seem especially diversified concerning technology. A comparison of major FBMs and a sample of electronics multinationals, for instance, shows a more differentiated technological structure in the former than in the latter (von Tunzelmann, 1998). While the coexistence of several scientific cultures could result in organization problems, diversification of technology could also provide the occasion for the cross-fertilization of technologies within the FBM (Granstrand and Sjölander, 1994), an important consideration given that the food industry depends, as stated, on innovation developed in upstream industries. The possibility of combining a variety of techniques could be one of the ownership advantages of the FBM versus most single-nation companies in this sector.

As noted, FBMs from different origins differ in multiple respects concerning the organization of R&D (location and type of laboratories, for instance). U.S., European, and Japanese companies also differ in patterns of patenting by field (von Tunzelmann, 1998). Although the share of food in total company patenting is similar in the three regions, in other technological fields the structure is very different. The share of biotechnology and chemical patents in Japanese firms is about twice as high as in the other two regions. Both fields together account for nearly two thirds of the innovative activities of Japanese FBMs. One reason could be that FBMs from other regions source their requirements for upstream technology from other companies, while the Japanese internalize them. This interpretation agrees with our previous observation on the organization of R&D in the Japanese FBMs. Such companies tend to develop, as stated, independent affiliates devoted to expanding generic knowledge, such as biotechnology, which are applicable to a variety of agribusinesses and food industries. In spite of such particularities, the shares of different fields by national groups of FBMs tended to remain stable from 1969-1979 to 1990-1994. No trend toward specialization of different regions could be detected (von Tunzelmann, 1998), suggesting that investing in a variety of technological fields could be strategically important per se for the FBM. Combining different types of innovation could be worthwhile too, as we show in the next subsection.

Technical and Design Innovation

FBMs combine technical and design innovation rather than using one type of innovation as a substitute for the other.

According to the U.S. Patent and Trademark Office (USPTO, 2003) a design consists of "the visual ornamental characteristics embodied in, or applied to, an article of manufacture." Designs may relate to the shape of an article, to its surface, or to a combination of both. A design patent protects "only the appearance of the article and not its structural or utilitarian features." While a utility patent protects "the way an article is used and works," a design patent protects the way it looks. As the USPTO Web page (www.uspto.gov/web/office) explains, minimal differences between similar designs can render each patentable. The legal protection given in the United States to utility and design patents is similar, except for the term, which is twenty years for the former and only fourteen for the latter.

FBMs register a small but increasing number of design patents. Of the total number of patents granted to the FBMs in our sample, 96 percent were, over 1969-1994, utility patents and only 4 percent design patents, 62 percent of them related to food packaging. With marketing and brand elements seemingly becoming increasingly important to FBMs, however, the number of their design patents rose even more quickly than their utility patents toward the end of the period (see Figure 3.1).

The production of utility and design patents is associated at the company level. Econometric analyses of time series of utility and design patents granted to the 103 world's largest FBMs over 1977-1994 show that such companies develop bundles of design and technical innovation, which are complementary (Alfranca, Rama, and von Tunzelmann, 2003b). We find similar percentages of R&D abroad for technical and design innovation, a finding which corroborates not only that multinationals transfer knowledge to foreign subsidiaries at more than one level (Archibugi and Iammarino, 2000) but also that technical and design innovation are complementary among FBMs.

The Forces Behind Innovation

Two of the forces that shape innovation in the multinational agro-food sector are self-generated accumulative innovation and competition.

The Persistence of Internal Research

As mentioned, the population of innovators is stable in the multinational agrofood sector. Moreover, among FBMs, past endogenous innovation is the most significant predictor of current innovation, much more so than other exogenous causes influencing innovation at the company level.

Panel-data analyses covering patents granted to the top group over 1977-1994 suggest that the multinational agrofood sector shows a stable pattern of technological accumulation in which "success breeds success" (Alfranca, Rama, and von Tunzelmann, 2002). In other words, the current population of inventive FBMs is chiefly made up of "old" innovators, large and small, who had produced patented inventions in the past. Steady flows of innovation among the heavy patentors in the multinational agrofood sector are another specificity of FBMs versus large companies in other sectors. In new and still undefined industries, where a great number of companies are still searching for engineering designs, other authors have instead observed short bursts of innovation followed by periods with no innovation, followed again by new waves of innovation (Clark, 1985). Unlike in many high-tech sectors, dominant FBMs are not systematically dislodged by new innovators, but innovation is cumulative *within* dominant firms.

Among FBMs, persistent innovators can combine both technical and design innovation to deter entry and could be better able than sporadic innovators to use innovation as a barrier to entry and mobility. Self-generated accumulative innovation is relevant among FBMs, but competition is also an ingredient of their technological dynamism This raises the next question.

Technological Competition

When new foreign entrants bring new technology to a domestic market controlled by established domestic companies, often multinationals themselves, the rules of competition could change (for a model of technological competition, see Eden and Molot, 1996). Even if incumbents do not initially detect that the foreign multinationals' market penetration is based on superior technology, or find it difficult to duplicate, in the medium run technological competition between both types of companies is likely to follow.

Among FBMs, opportunities for interacting strategically are frequent because such companies coexist in the most important markets of North America and the European Union, and sometimes in large developing countries (Alfranca, Rama, and von Tunzelmann, 2003). FBMs compete not only in price and marketing techniques but also in new technology and brands, an element influenced by packaging design.

In a sample of major FBMs to which nearly 17,000 patents were granted over 1977-1994, companies built chiefly on their own past innovation and their design experience (Alfranca, Rama, and von Tunzelmann, 2003). They also seem to react, however, with a short lag, to the innovative activities of other FBMs operating in the same subsector, i.e., agribusiness and basic food, beverages, and highly processed food. Pressures from rivals and their stimulating effects at the company level are perceptible not only in the technical field but also concerning instruments of differentiation of products, such as design of food packaging.

CONCLUDING REMARKS

In this chapter, we have studied the characteristics of innovation in food and beverage multinationals. Since the early 1980s, the number of patents granted to such firms in the United States has risen. Over 1969-1994, European and U.S. companies contributed a larger number of innovations than Japanese FBMs. However, Japanese firms were more dynamic and conceived a growing number of inventions. Among American FBMs, the pace of innovation lost momentum. By contrast, Japanese companies and firms based in EU countries that were not traditional sources for FDI started to catch up in technological development. Divergent technological trends between U.S. FBMs and other FBMs coincided with more general trends in international businesses. U.S. multinationals have tended to lose ground to companies based in other countries, both in the F&B industry (Chapter 1) and in other international businesses (Dunning, 1993). Another, more speculative, explanation for changes in the top group of FBMs is that behind the recent emergence of second-tier firms and new source countries in F&B is their innovation during the recent period. In the future, Japanese FBMs could play an increasing role in food and food-related innovation, owing to their technological dynamism and their R&D organization, which favors generic technology (for instance, fermentation processes) and interdisciplinary knowledge applicable to many food and biotechnological fields.

Compared to multinationals in other sectors, FBMs in our sample locate a large part of their R&D activities abroad, which confirms previous analyses (Cantwell and Hodson, 1991; Cantwell and Janne, 2000; Patel and Pavitt, 1991). As FBMs also seem to internationalize a larger part of their productive facilities than multinationals in other sectors (Stopford and Dunning, 1983), such results also confirm that multinationals tend to disperse their R&D outlays abroad when the foreign sales of their subsidiaries are important (Zejan, 1990). FBMs' extensive internationalization of research activities could be attributed not only to adaptive tasks, as traditionally argued, but also to more independent and creative functions.

It should be stressed that even if FBMs research abroad more than other multinationals, most of their innovative activities are still generated at home. In this respect, however, differences among national groups of companies are enormous. U.S. FBMs locate a *small* though growing part of their research in foreign nations, especially in Europe. European FBMs, instead, locate *most* of their research abroad. As for many other multinationals (Rugman and Verbeke, 1995), they could be currently moving toward a structure of multiple home bases concerning their innovative activities so as to improve their international competitiveness.

This is not a regionalization process, as often held. Most of this foreign research is located not within Europe, as might be expected, but in the United States. FBMs are the European multinationals that locate a particularly large share of their innovative activities in the United States. This behavior could reflect inadequacies in the European systems of innovation related to the agrofood sector, a matter of concern given the importance of this industry for EU trade and FDI. It could also indicate insufficient supranational linkages within the European Union, in projects dealing with food technology or new key areas related to agrofood production, such as biotechnology. More investigation is needed on the content of European FBMs' research in the United States in order to understand this phenomenon better.

Each FBM is likely to enjoy a specific combination of skills and technical advantage. First, FBMs combine technical and design innovation rather than using one type of innovation as a substitute for the other. This contradicts common wisdom supposing that food firms innovate in packaging and presentation of products as an alternative to intrinsic innovation in products and processes. Furthermore, the interconnectedness of technical and design innovation at the company level is probably a barrier to entry in the multinational agrofood industry. Second, FBMs innovate in both food and nonfood fields. They assign a substantial part of their efforts to nonfood not only because they are diversified into noncore products but also, and more important, because they intend to master the upstream technology that is essential in F&B processing (Christensen, Rama, and von Tunzelmann, 1996; Rama, 1996b). The possibility to combine a variety of innovations could be an ownership advantage of FBMs versus one-nation companies.

Among FBMs, the production of patented inventions is closely associated with past research efforts in the multinational itself. In this multinational sector, "success breeds success." Knowledge is cumulative. Dominant FBMs are not dislodged by "new" innovators, as happens with large companies in some other sectors. Here the pattern of innovation is one of "creative accumulation" rather than "creative destruction" (Cefis and Orsenigo, 2001). In the literature on patent races, very large companies tend to be "old" innovators but not current innovators because they fear "cannibalizing" their own previous inventions (Geroski, Van Reenen, and Walters, 1997). Among FBMs, by contrast, those that currently foster both technical changes and new designs of packaging are "old" innovators.

Though many giant food firms innovate only sporadically, to be permanently innovative could be relevant for FBMs because strategic asset *stocks,* such as the stock of knowledge, are accumulated over periods of time. Brands have similar cumulative properties since they "need maintenance" (Telser, 1961), including improvements in packaging design. Thus occasional innovators are less likely to maintain the advantages derived from their new products, processes, or designs, especially because, in this industry, imitation is relatively easy (OECD, 1988).

In addition to being positively stimulated by their own innovative history, FBMs also respond to competitive pressures from rival innovators who operate in the same subsector within this industry.

Compared to other multinationals, FBMs are an outstanding example of technoglobalism. FBMs tend to locate abroad a larger share of their foreign activities. In other multinationals, the foreign share of research is likely to remain stable; in FBMs it tends to grow quickly. When investing in R&D abroad, other European multinationals tend to invest within Europe; as mentioned previously, European FBMs tend to innovate in the United States. The FBM shows other specificities vis-à-vis other multinationals. The available evidence suggests that FBMs could be more technologically diversified and devote more attention to innovation in packaging design.

FBMs remain the world's most important players concerning food technology but are indeed losing ground to other innovators. Related to this, a number of developments suggest that, in spite of the importance of FBMs, national systems of innovation still have an important role to play in the agrofood sector. First, although the multinational enterprise has always been viewed as a source of new knowledge (Caves, 1996), not all FBMs are innovative and many innovate only sporadically. Second, the majority of FBMs still locate most of their R&D activities in their respective home countries. Governments from developing countries should be aware of such limitations on innovation in FBMs and carefully select investors. It is often believed that FDI will automatically bring state-of-the-art technology to the host country, but some are not innovative at all. On the other hand, national systems of innovation could be a stimulus for the internationalization of domestic firms in this industry. In the home country, linkages and R&D spillovers within the food and drink industry as well as between F&B, upstream sectors, research centers, and universities could aid food and drink firms to go international.

APPENDIX

The Data

The Company Sample

We selected eighty-three continuing companies, i.e., firms included in the top group over 1969-1994, from AGRODATA (IAMM, 1990; Padilla et al., 1983; Rastoin et al., 1998). Produced by the Institut Agronomique Méditerranéen de Montpellier (France) (IAMM), this database gathers information on the world's 100 largest food multinationals since the 1970s. The data sources are *Moody's Industrial Manual*, the Fortune Directory of the 500 largest U.S. and the 500 largest non-U.S. corporations, the "Dossier 5,000" of the largest European companies published by *Le Nouvel Economiste*, Dun & Bradstreet reports, and the annual reports of the enterprises, among others.

The Patent Sample

The variable used here is the number of patented innovations granted to the firms in our sample in the United States over 1969-1994. The Science and Technology Policy Research (SPRU) at the University of Sussex (United Kingdom) collected the patent data from the U.S. Patent and Trademark Office (USPTO). The data from 1975 onward (only) are available online from the USPTO <http://www.uspto.gov>. However, working the data from online sources or CD-ROM into usable results still involves intensive research efforts. Basically, the USPTO assignees are given according to the name of the organization to which they are directly affiliated, rather than the name of the corporation. A large patenting firm, such as Unilever, may have hundreds of these patenting subunits in addition to the core corporation, and the task of consolidating them into corporate totals is a major one, since the USPTO database does not record their ownership. The latter information has to be painstakingly constructed from sources (such as *Who Owns Whom?*) before searching the database and then aggregating the number of patents. The portions of the FBM's innovations generated at home and offshore were calculated.

"Food" patents cover the three-digit classes of the USPTO as follows: 426 ("food or edible material: processes, compositions, and products"), 127 ("sugar, starch, and carbohydrates"), and 099 ("food and beverages: apparatus"). Tobacco patents are from class 131. A full concordance with the 400-odd USPTO classifications is too long to publish here but is available from N. von Tunzelmann on request.

To analyze the FBM patenting at home and abroad, we use data pertaining to laboratories located in Argentina, Australia, Canada, Denmark, France, Italy, Japan, South Africa, Sweden, Switzerland, the Netherlands, the United Kingdom, and the United States. Though FBMs also generate patents from laboratories in thirty-one other host countries, the patents analyzed here account for around 85 percent the total number of patents granted to the companies in our sample.

Product and Process Innovation

It is unfortunately not easily possible to distinguish process from product patents by using the U.S. classification system, short of detailed scrutiny of each patent specification, which for the more than 16,400 patents covered here is beyond our means.

There are two reasons for this. The first is that the U.S. classification mixes product and process patents at the three-digit level, as the titles of the classes indeed indicate (see the previous example for class 426). The second is the inherent technical difficulty of deciding whether a particular patent refers to a product or a process innovation, e.g., for many F&B patents in the area of ingredients or chemicals—within a single company it may happen that a product innovation is developed at one site to be used as a process innovation on another site (and so on).

NOTES

1. There are some reasons for this perception. Of the nearly 2 million patents granted by the U.S. Patent and Trademark Office (USPTO) from 1969 to 1994, only 23,022 were food patents, including food proper, F&B apparatus, and tobacco (Christensen, Rama, and von Tunzelmann, 1996). Moreover, the growth of food patents was less dynamic. They grew only 8.37 percent between 1969-1973 and 1990-1994, while patents in all sectors increased 35.07 percent, though allowances should be made on the grounds that food firms have a lower propensity to patent their inventions than companies in other industries (Scherer, 1984). The relatively small ratio of R&D expenditures to sales in F&B also suggests that F&B is, relatively, a low-tech industry. In the European Union (EU), for instance, reported ratios for large firms were 0.5 percent in food manufacturing versus, for instance, 12 percent in drugs (Grunert et al., 1995). Nevertheless, recent studies are beginning to question the perception of F&B as a low-tech industry, basically because the sector is actually at the forefront of industries in the *application* of a breadth of different scientific and technological advances (Christensen, Rama, and von Tunzelmann, 1996; von Tunzelmann, 1998).

2. For instance, many successful inventions are never patented (Rosenberg, 1982), and firms from different countries or industries show different propensities to

patent (Archibugi and Pianta, 1992). Moreover, patent counts give no information on the technical importance or the market value of the inventions.

3. The multinationals in their sample held around 49.1 percent of the patents in thirty-three industries.

4. The source and the methodology for obtaining these data is similar to that used for the FBMs' patents (see the Appendix). One of the reasons for faster growth of patenting among nonmultinational inventors in this industry could be that smaller foreign firms now have a greater propensity than before to patent in the United States. By contrast, foreign FBMs are less likely to have changed their propensity to patent in the United States, since their interests there have a longer history. However, such differences between multinationals and nonmultinational inventors could also occur in nonfood sectors, so this factor is unable to explain the falling FBM share.

5. Allowances should be made, however, for the fact that not all biotech patents granted to non-FBMs are used in F&B. Non-FBMs that are granted biotech patents include multinationals and other companies in nonfood sectors, e.g., drugs. In part, differences among FBMs and other inventors could be attributable to different technological opportunity among sectors.

6. The source and methodology for gathering the data are described in the Appendix.

7. Their calculations are also based on U.S. patents and their study period, 1969-1995, is similar to ours.

8. The column concerning multinationals should be compared to the utility patents in FBMs, since Cantwell and Janne's database includes utility patents only, not design patents.

9. Even some giant firms, such as Archer Daniels Midland Co. (United States), Associated British Foods (United Kingdom), Koninklijke Wessanen (Netherlands), Molson Co. (Canada), or Union Laitière Normande (France), patented small numbers of inventions in the United States.

REFERENCES

Acs Z. J. and Audretsch D. B. (1989). Patents as a measure of innovative activity. *WZB, Berlin:* 1-13.

Alfranca O., Rama R., and von Tunzelmann N. (2001). Cumulative innovation in food and beverage multinationals. In D. Kantarelis (Ed.), *Business and Economics Society International Conference 2001: Proceedings—Global business and economics review, anthology 2001* (pp. 446-459). Worcester, MA: BESI.

Alfranca O., Rama R., and von Tunzelmann N. (2002). A patent analysis of global food and beverage firms: The persistence of innovation. *Agribusiness: An International Journal,* 18(3) 349-368.

Alfranca O., Rama R., and von Tunzelmann N. (2003). Competitive behaviour, design and technical innovation in food and beverage multinationals. *International Journal of Biotechnology/International Journal of Technology Management* (Special issue on "Innovation in Food and Beverages, and Biotechnology"), 5(3/4): 222-248.

Alfranca O., Rama R., and von Tunzelmann G. N. (2004). Innovation spells in the multinational agri-food sector. *Technovation*, 24(8): 597-672.

Anastassopoulos G., Papanastassiou M., Pearce R. D., and Traill W. B. (1997). Firm and location: Specific determinants in investment and trade strategies of major multinationals in the food industry in Europe. In S. R. Henneberry (Ed.), *Foreign direct investment and processed food trade* (pp. 57-78). Stillwater, Oklahoma: State University.

Archibugi D. and Iammarino S. (2000). Innovation and globalization. In F. Chesnais, G. Ietto-Gillies, and R. Simonetti (Eds.), *European integration and global strategies* (pp. 95-120). London and New York: Routledge.

Archibugi D. and Michie J. (1995). The globalisation of technology: A new taxonomy. *Cambridge Journal of Economics*, 19: 121-140.

Archibugi D. and Pianta M. (1992). Specialization and size of technological activities in industrial countries: The analysis of patent data. *Research Policy*, 21: 79-93.

Bijman, W.B., van Tulder R., van Vliet M. (1997). Internationalisation of Dutch agribusiness and the organisation of R&D. Paper presented at the EAAE Seminar on Globalization of the Food Industry: Policy Implications, Reading, UK, Centre for Food Economics Research.

Bound J., Cummins C., Griliches Z., Hall B. H., and Jaffe A. (1984). Who does R&D and who patents? In Z. Griliches (Ed.), *R&D, patents, and productivity* (pp. 21-54). Chicago: University of Chicago Press.

Cantwell J. (1989). *Technological innovation and multinational corporations*. Oxford and Cambridge: Basil Blackwell.

Cantwell J. and Hodson C. (1991). Global R&D and UK competitiveness. In M. Casson (Ed.), *Global research strategy and international competitiveness* (pp. 133-183). Oxford and Cambridge: Basil Blackwell.

Cantwell J. and Janne O. (1999). Technological globalization and innovative centres: The role of corporate technological leadership and locational hierarchy. *Research Policy*, 28(2-3): 119-144.

Cantwell J. and Janne O. (2000). Globalization of innovatory capacity: The structure of competence accumulation in European home and host countries. In F. Chesnais, G. Ietto-Gillies, and R. Simonetti (Eds.), *European integration and global corporate strategies* (pp. 121-177). London and New York: Routledge.

Caves R. E. (1996). *Multinational enterprise and economic analysis* (Second edition). Cambridge, UK: Cambridge University Press.

Cefis E. and Orsenigo L. (2001). The persistence of innovative activities: A cross-countries and cross-sectors comparative analysis. *Research Policy*, 30: 1139-1158.

Christensen J. L., Rama R., and von Tunzelmann N. (1996). Study on innovation in the European food products and beverages industry. Monograph number 145. The European Commission: EIMS/SPRINT Brussels.

Clark K. B. (1985). The interaction of design hierarchies and market concepts in technological evolution. *Research Policy*, 14(5): 235-251.

Connor J. M. (1981). Food product proliferation: A market structure analysis. *American Journal of Agricultural Economics,* 10(1): 25-52.

Connor J. M. (1997). Economic overview and research issues: On the convergence of food systems. *Agribusiness,* 13(2): 253-259.

Criscuolo P., Narula R., and Verspagen B. (2002). The relative importance of home and host innovation systems in the internationalisation of MNE R&D: A patent citation analysis. MERIT-Infonomics Research Memorandum series, No. 26. Maastricht, the Netherlands: Universiteit Maastricht.

Dunning J. H. (1993). *The globalisation of business.* London and New York: Routledge.

Dunning J. H. (1994). Multinational enterprises and the globalization of innovatory capacity. *Research Policy,* 23: 67-68.

Eden E. and Molot M. A. (1996). Made in America? The US auto industry, 1955-95. *The International Executive,* 38 (4): 501-541.

Fanfani R., Lanini L., and Torroni S. (1996). Invention patents in Italian agro-food industry: Analysis of the period 1967-1990. In G. Galizzi and L. Venturini (Eds.), *Economics of innovation: The case of the food industry* (pp. 391-406). Heidelberg: Physica-Verlag.

Freeman C. (1994). The economics of technical change. *Cambridge Journal of Economics,* 18: 463-514.

Geroski P. A., Van Reenen J., and Walters C. F. (1997). How persistently do firms innovate? *Research Policy,* 26: 33-48.

GEST (1986). *Grappes technologiques: Les nouvelles stratégies d'entreprise.* Paris: McGraw-Hill.

Gonard T., Green R. H., Malerbe A., and Requillart V. (1991). Changement technique et stratégie des acteurs dans le secteur de la chimie du sucre [Technical change and strategy in the sugar processing industry]. *INRA, Economie et Sociologie Rurales* (Special issue on "Changement technique et restructuration de l'industrie agro-alimentarie en Europe" [Technical Change and Reorganization of the European Agrofood Industry]), 7: 143-158.

Granstrand O., Hakansson H., and Sjölander S. (1993). Internationalization of R&D—A survey of some recent research. *Research Policy,* 22: 413-430.

Granstrand O. and Sjölander S. (1994). Managing innovation in multi-technology corporations. In M. Dodgson and R. Rothwell (Eds.), *The handbook of industrial innovation* (pp. 367-383). Cheltenham, UK: Edward Elgar.

Grunert K. G., Harmsen H., Meulenberg M., Kuiper E., Ottowitz, T., Declerck F., Traill B., and Göransson G. (1995). Innovation in the food sector: Between technology-push and demand-pull. Structural change in the European food industries (within the EU AAIR programme). European Union discussion paper number 10 (December).

Hakanson L. and Nobel R. (1993). Determinants of foreign R&D in Swedish multinationals. *Research Policy,* 22: 397-411.

Hu Y.-S. (1992). Global or stateless corporations are national firms with international operations. *California Management Review,* 34(Winter): 107-126.

Institut Agronomique Méditerranéen de Montpellier (IAMM) (1990). *Les 100 premiers groupes agro-alimentaires mondiaux.* Montpellier, France.

Iussaume R. A. Jr. and Kenney M. (1993). Japanese investment in United States food and agriculture: Evidence from California and Washington. *Agribusiness,* 9(4): 413-424.

Khanna T. and Singh J. (2002). What drives innovation by foreign multinationals? Harvard Business School Strategy Unit working paper number 03-058: 43, Boston, MA.

Kuemmerle W. (1999). Foreign direct investment in industrial research in the pharmaceutical and electronics industries—Results from a survey of multinationals. *Research Policy,* 28: 179-193.

National Science Foundation (1996). *Science and engineering indicators 1996.* Washington, DC: U.S. Government Printing Office.

Nicolas F. (1996). Combined roles of process and product innovation in the food industries. In G. Galizzi and L. Venturini (Eds.), *Economics of innovation: The case of the food industry* (pp. 341-353). Heidelberg: Physica-Verlag.

Organisation for Economic Co-operation and Development (OECD) (1979). *Impact of multinational enterprises on national scientific and technical capacities.* Paris: OECD.

OECD (1988*). Industrial revival through technology.* Paris: OECD.

Padilla M., Laval G. G., Allaya M.-C., and Allaya M. (1983). *Les cent premiers groupes Agro-Industriels Mondiaux.* Montpellier, France: IAMM.

Patel P. (1995). Localised production of technology for global markets. *Cambridge Journal of Economics,* 19: 141-153.

Patel P. and Pavitt K. (1991). Large firms in the production of the world's technology: An important case of "non-globalisation." *Journal of International Business Studies,* 22: 1-21.

Patel P. and Pavitt K. (1997). The technological competencies of the world's largest firms: Complex and path-dependent, but not much variety. *Research Policy,* 26: 141-156.

Patel P. and Vega M. (1999). Patterns of internationalisation of corporate technology: Location vs. home country advantages. *Research Policy,* 28: 145-155.

Pavitt K. (1984). Patterns of technical change: Toward a taxonomy and a theory. *Research Policy,* 13: 343-373.

Pearce R. (1999). The evolution of technology in multinational enterprises: The role of creative subsidiaries. *International Business Review,* 8: 125-148.

Rama R. (1991). El entorno tecnológico de la empresa alimentaria. *INRA Economie et Sociologie Rurales, Paris* 7(Special issue: Changement technique et restructuration de l'industrie agro-alimentaire en Europe): 59-93.

Rama R. (1996a). An empirical study on sources of innovation in the international food and beverage industry. *Agribusiness: An International Journal,* 12: 123-134.

Rama R. (1996b). Les multinationales et l'innovation: Localisation des activités technologiques de l'agro-alimentaire. *Economie Rurale, Paris* 231(January/February): 62-68.

Rama R. (1999). Innovation and profitability of global food firms: Testing for differences in the influence of the home base. *Environment and Planning, A* 31: 735-751.

Rastoin J.-L., Ghersi G., Pérez R., and Tozanli S. (1998). *Structures, performances et stratégies des groupes agro-alimentaires multinationaux.* Montpellier, France: AGRODATA.

Rastoin J.-L. and Tozanli S. (1998). Géostratégies des firmes agroalimentaires. In J.-L. Rastoin (Ed.), *Mondialisation et géostratégies agroalimentaires,* Volume 109 (pp. 51-67). Montpellier, France: AGRO-INRA.

Reddy, P. (1993). Emerging patterns of internationalization of corporate R&D: Opportunities for developing countries? In C. Brundenius and B. Göranson (Eds.), *New technologies and global restructuring: The third world at a crossroads.* London: Taylor Graham.

Reddy P. (2000). *Globalization of corporate R&D: Implications for innovation systems in host countries.* London and New York: Routledge.

Rosenberg N. (1982). *Inside the black box: Technology and economics:* Cambridge: Cambridge University Press.

Rugman A. E. (1987). The firm specific advantages of Canadian multinationals. *Journal of International Economic Studies,* 2: 1-14.

Rugman A. E. and Verbeke A. (1995). Transnational networks and global competition. An organizing framework. *Research in Global Strategic Management,* 5: 3-23.

Scherer F. M. (1984). *Innovation and growth: Schumpeterian perspectives.* Cambridge, MA: MIT Press.

Soete L. (1987). The impact of technological innovation on international trade patterns: The evidence reconsidered. *Research Policy,* 16: 101-130.

Stopford J. M. and Dunning J. H. (1983). *Multinationals: Company performance and global trends.* London: McMillan Publishers.

Telser L. (1961). How much does it pay whom to advertise? *American Economic Review (Proc.),* 51(2): 194-205.

von Tunzelmann G. N. (1998). Localized technological search and multi-technology companies. *Economics of Innovation and New Technology,* 6: 231-255.

Wilkinson J. (1998). The R&D priorities of leading food firms and long-term innovation in the agrofood system. *International Journal of Technology Management,* 16: 711-720.

Zanfei A. (2000). Transnational firms and the changing organisation of innovative activities. *Cambridge Journal of Economics,* 24: 515-542.

Zejan, M.C (1990). R&D activities in affiliates of Swedish multinational enterprises. *Scandinavian Journal of Economics,* 92: 487-500.

Chapter 4

Foreign Direct Investment in U.S. Food and Kindred Products

Daniel Pick
Thomas Worth

INTRODUCTION

The composition of international trade and its structure changed considerably during the 1990s. While trade in bulk commodities was central in terms of policy focus as well as research focus, the fact is that global commerce in processed food products significantly exceeds trade in bulk commodities. For example, U.S. exports of processed food products in 1998 amounted to $29.4 billion, compared to exports of $8.7 billion in bulk commodities.

While trade is an important channel of marketing U.S. food products, the globalization of international markets introduced various marketing vehicles to accommodate the increasing international demand for processed food products. Such vehicles include foreign production by U.S.-owned multinational enterprises (MNEs) as well as other contractual activities. These different activities include investments by home-country firms in the production activities in host countries. Such investment activities are referred to as foreign direct investment (FDI).

The role of FDI in the globalization of food markets has implications for the agricultural sector in various ways. The purpose of this

Daniel Pick is the branch chief and Thomas Worth is an economist in the Specialty Crops Branch, the U.S. Department of Agriculture, Economic Research Service. The views expressed here are the authors' and do not necessarily represent those of the Department of Agriculture.

chapter is twofold. First is to familiarize the readers with the importance of FDI to the agricultural sector. This will be done by defining what we mean by foreign direct investment and then illustrating the historical trends of FDI in the food-processing sector. A second purpose is to outline the different issues that are central to the relationships between foreign direct investment and the agricultural sector. We will then tie those issues to the existing empirical studies that link FDI to the food sector and discuss other issues that have not been dealt with yet with regard to FDI.

RECENT TRENDS IN FDI

The growth in FDI far outpaces growth in exports. As shown in Figure 4.1, the volume of FDI for all industries increased fivefold from 1981 to 1996, whereas trade volume tripled during this time. The trend is similar for FDI in agriculture-related industries. The majority of FDI flows consist of mergers or acquisitions of existing foreign assets (see Chapter 1). The rapid increase in FDI flows is due to deregulation, privatization, and trade policy reforms that many countries have pursued since the early 1990s.

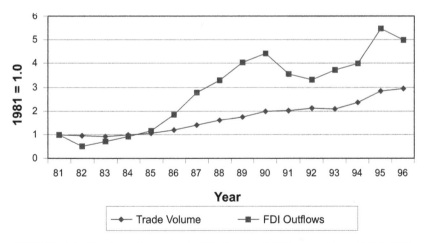

FIGURE 4.1. Trade volume and FDI flows, OECD countries, 1981-1996. (*Source: International Direct Investment Yearbook,* OECD, various years; *Yearbook of International Statistics,* United Nations, various years.)

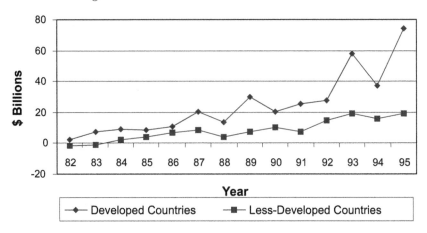

FIGURE 4.2. FDI outflows from the United States, 1982-1996. (*Source: International Direct Investment Yearbook,* OECD, varous years.)

The increase in FDI flows is not distributed evenly around the world. Figure 4.2 shows the destination of FDI from all U.S. industries. Most FDI from the United States flows to other developed countries. This trend is present in both agricultural and nonagricultural industries. In general most FDI flows in the world both originate from and are destined for developed countries. The gap between FDI bound for developed countries versus FDI bound for less-developed countries has widened throughout the 1980s and 1990s.

Since FDI represents ownership (or at least partial ownership) and control over foreign production, another useful measure of the magnitude and importance of FDI is the value of sales of affiliates of multinational companies. Figure 4.3 shows the sales of U.S. affiliates of foreign firms for several agriculture-related industries as well as all imports of processed food products. Sales by U.S. affiliates of foreign firms in the processed food industry (food and kindred products) reflect the trend in FDI flows for the world. Affiliate sales have grown more quickly than have imports and now far exceed foreign trade volumes. In 1999, for example, U.S. affiliates of foreign firms had $60 billion in sales, whereas imports of similar goods was only $27 billion.

U.S. multinationals have invested in other food-related industries. One of them is the retail trade industry, which consists mainly of food

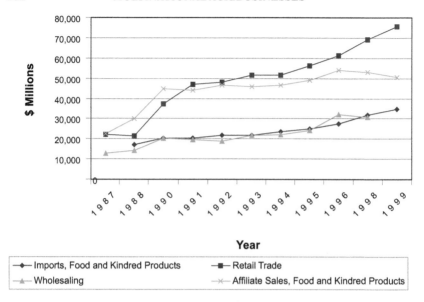

Year

◆ Imports, Food and Kindred Products	■ Retail Trade
▲ Wholesaling	✕ Affiliate Sales, Food and Kindred Products

FIGURE 4.3. Sales by U.S. affiliates of foreign firms, 1987-1999. (*Source: Survey of Current Business,* U.S. Department of Commerce, Bureau of Economic Analysis, varous issues.)

stores. Sales from U.S. affiliates of foreign firms in retail trade have grown at a pace similar to the processed food industry, nearly tripling from 1987 to 1999. Another industry in which foreign multinationals have acquired U.S. assets is wholesale trade. The growth in affiliate sales in the wholesale trade industry, while significant, is more gradual than in the processed food and retail trade industries. Added together, sales of food-related foreign affiliates dwarf food imports into the United States.

The U.S. affiliates of foreign multinationals in the food-processing industry produce a variety of products. The products and their share of foreign affiliate sales are shown in Figure 4.4. The single largest category is beverages and tobacco products (25 percent) followed by grain and oilseed milling (13 percent) and dairy products (12 percent). Multinational investments in these products may reflect the fact that they are difficult or expensive to export over long distances. The presence of a production facility near the demand market enables

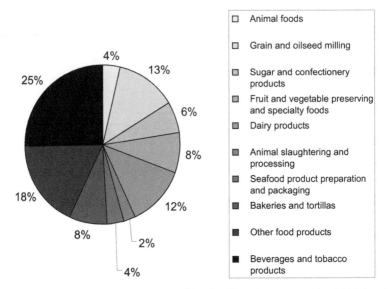

FIGURE 4.4. Sales by U.S. affiliates of foreign firms by commodity, 1999 (total = $69.8 billion). (*Source:* Operations of U.S. Affiliates of Foreign Companies, U.S. Department of Commerce, Bureau of Economic Analysis, 1999.)

MNEs to supply output at a lower cost than by exports from their home countries.

Most U.S. affiliate sales are from firms located in other developed countries. As shown in Figure 4.5, firms from Europe and Canada account for more than two-thirds of U.S. affiliate sales in food and kindred products. As of 1999, the United Kingdom had the most foreign affiliate sales in the United States of any of the European countries. This may reflect the effect of its cultural similarity with the United States. After the United Kingdom, Germany and the Netherlands had the next most affiliate sales. In Asia, Australia and Japan account for most of the foreign affiliate sales. The developing countries with the most foreign affiliate sales are from Latin America, with Mexico claiming the largest share, followed by Brazil and Argentina.

The distributions of affiliate sales in the United States are different for the retail trade and wholesale trade sectors. For the retail trade sector, as shown in Figure 4.6, firms based in Europe claim 84 percent of affiliate sales. This reflects the acquisition of food stores by Euro-

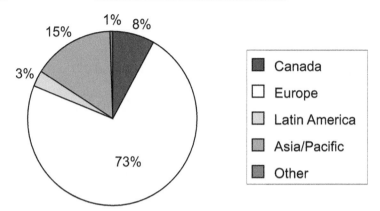

FIGURE 4.5. Sales by U.S. affiliates of foreign firms, food and kindred products, 1999 (total $50.4 billion). (*Source:* Operations of U.S. Affiliates of Foreign Companies, U.S. Department of Commerce, Bureau of Economic Analysis, 1999.)

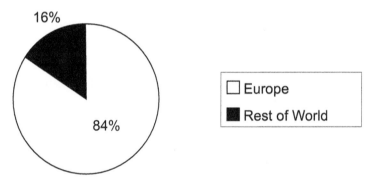

FIGURE 4.6. Sales by U.S. affiliates of foreign firms, retail trade, by parent company origin, 1999 (total = $75.7 billion). (*Source:* Operations of U.S. Affiliates of Foreign Companies, U.S. Department of Commerce, Bureau of Economic Analysis, 1999.)

pean companies. The distribution of affiliate sales in wholesale trade, shown in Figure 4.7, has a more even geographic distribution.

Outflows of FDI from the United States to the rest of the world show a similar pattern to FDI inflows. Most FDI is bound for other developed countries. The geographic distribution of the sales of foreign affiliates of U.S. firms in the food and kindred products industry

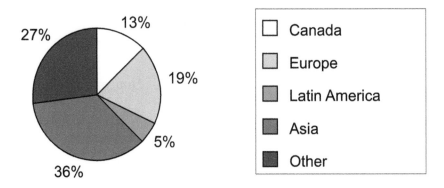

FIGURE 4.7. Sales by U.S. affiliates of foreign firms, wholesale trade (groceries), by parent company origin, 1996 (total = $32.4 billion). (*Source:* Survey of Current Business, U.S. Department of Commerce, Bureau of Economic Analysis, 1996.)

is shown in Figure 4.8. European firms are responsible for a majority of foreign affiliate output in the United States, as Figure 4.5 shows. Similarly, the majority of foreign affiliate sales of U.S. firms takes place in Europe. Within Europe, more than half of the affiliate sales take place in the United Kingdom, Germany, and the Netherlands. The largest single location for foreign affiliate sales is Canada. This is partly due to some of the production in Canada being exported into the United States. Asia and Latin America have nearly equal shares of foreign affiliate sales. The developing countries with the largest share of U.S. foreign affiliate sales are Brazil (see Chapter 9) and Mexico.

Although both inbound and outbound FDI in food and kindred products has the same geographical distribution, the magnitude is different. In 1996, the value of foreign affiliate sales of U.S. firms, at around $100 billion, was double the value of the sales of U.S. affiliates of foreign firms. This is also reflected in the number and size of multinational food-processing firms. Table 4.1 lists the world's twenty-five largest food-processing firms. Fourteen of the twenty-five are based in the United States. The rest are in Europe or Japan.

MULTINATIONAL AGRIBUSINESSES

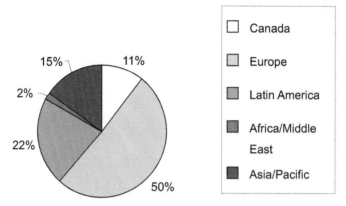

FIGURE 4.8. Sales by foreign affiliates of U.S. firms, food and kindred products, 1998 (total = $133.1 billion). (*Source:* Operations of U.S. Parent Companies and Their Foreign Affiliates, 1998, U.S. Department of Commerce, Bureau of Economic Analysis, 1996.)

WHAT IS FOREIGN DIRECT INVESTMENT AND WHY?

International trade is often associated with the transfer of goods and services from one country to another. However, the theory of international trade also places extreme importance on the movements of factors of production across countries. Factor movements can take the form of labor migration across borders or transfer of capital through international borrowing and lending or through the creation of multinational enterprises.

International lending and borrowing differ from foreign direct investment in that no commitments are made by the borrower but the commitment to repay the loan. In foreign direct investment, we refer to the case of cross-border capital flows, in which a firm creates or acquires control of a subsidiary in another country. The subsidiary does not have an obligation to repay any part of the capital flow since it has become part of the firm.

Most of the world's FDI is funded through multinational corporations who provide the necessary financing to control operation of subsidiaries. The theory of multinational enterprises provides us with two reasons as to why foreign direct investment through multina-

TABLE 4.1. Country of headquarters and sales of the world's 25 largest food-processing firms, 1993.

Company	Headquarters	Processed food sales ($ billion)	Total company ($ billion)
1. Nestle	Switzerland	36.3	39.1
2. Phillip Morris/Kraft	United States	33.8	50.6
3. Unilever	United Kingdom/ Netherlands	21.6	41.9
4. ConAgra	United States	18.7	23.5
5. Cargill	United States	16.7	47.1
6. PepsiCo	United States	15.7	25.0
7. Coca-Cola	United States	13.9	14.0
8. Dannon S.A.	France	12.3	12.3
9. Kirin Brewery	Japan	12.1	12.1
10. IBP, Inc.	United States	11.2	11.7
11. Mars, Inc.	United States	11.1	12.0
12. Anheuser-Busch	United States	10.8	11.5
13. Montedisor/Feruzzi/Eridania	Italy	9.9	12.3
14. Grand Metropolitan	United Kingdom	9.9	11.2
15. Archer Daniels Midland Co.	United States	8.9	11.4
16. Sara Lee	United States	7.6	15.5
17. Allied Domecq Plc	United Kingdom	7.2	7.2
18. RJR Nabisco	United States	7.0	15.1
19. Guinness Plc.	United Kingdom	7.0	7.0
20. H.J. Heinz	United States	6.8	7.0
21. Asahi Breweries	Japan	6.8	6.8
22. CPC International	United States	6.7	6.7
23. Dalgety	United Kingdom	6.7	6.7
24. Campbell Soup	United States	6.6	6.6
25. Bass Plc.	United Kingdom	6.6	6.6

Source: Henderson, Dennis R., Charles R. Handy, and Steven A. Neff (1996). *Globalization of the processed foods market.* Agricultural Economic Report No. 742, Economic Research Service, U.S. Department of Agriculture.

tional corporations takes place. First is the issue of location, which helps explain why production occurs in different countries by the same firm. Trade barriers and transportation costs are two possible explanations that have been mentioned as to why foreign direct investments and multinational activities occur. The second reason relates to the concept of internalization. Firms can increase their efficiency and their profitability by investing in foreign subsidiaries and conducting transactions within firm rather than between firms. This is particularly evident in the case of technology transfer and vertical integration.

Two types of foreign direct investment are often mentioned: outbound and inbound. The former refers to investment by home-based multinational firms in production and marketing facilities in foreign countries. The latter indicates investment by foreign firms in production and marketing facilities in the domestic or host country.

DETERMINANTS OF FDI

Various hypotheses have emerged to explain firms' strategic behavior regarding exports and foreign investment. These hypotheses have been developed in the industrial organization (IO) literature and empirically tested in various studies covering the manufacturing sector and, to a lesser degree, the food sector. The various variables that determine the firm's decision of whether to invest abroad or to directly export depend on factors associated with conditions in both the domestic market and the foreign market.

Host country variables affecting FDI include trade and investment policies, market size, market characteristics, and others. Domestic market variables determining FDI include intellectual property rights, economies of scale, and product differentiation. FDI in the agricultural industries (generally processed foods and related products) claimed 6 percent of total U.S. FDI in the manufacturing industries in 1996. Agricultural FDI follows similar patterns to other manufacturing industries. The agricultural industries are capital intensive and engage in FDI (rather than licensing) to maintain quality, protect a trademark, and take advantage of economies of scale (see Dunning, 1994). The majority of U.S. FDI in the food and agricultural industries is bound for Europe.

Intellectual Property Rights

The role of firms' intellectual property rights in the decision process of whether to invest abroad has been developed in the theoretical literature and only recently tested empirically. Grubaugh (1987) linked the theory of foreign direct investment to IO characteristics. Such linkage was made with the assumption that domestic firms develop foreign operations in host countries in order to exploit intangible assets (firm-specific assets) such as patents, trademarks, and research and development (Connor, 1983). These firm-specific assets, including intellectual property rights, are not easily priced and transacted, and firms prefer to invest in operations abroad. Thus the domestic firms will realize potential profits associated with such assets.

The role of firm-specific assets in facilitating investment abroad, as developed in the theoretical literature, was explored empirically by Lee and Mansfield (1996). The tested hypothesis was that countries' protection of intellectual property rights is directly related to FDI in these countries. The measure of intellectual property protection in a country is based on the results of a survey of multinational firms. A least squares estimate shows a positive relationship between strength of intellectual property protection and volume of U.S. FDI.

The evidence with respect to the food industry is almost nonexistent. One exception is a study by Henderson, Vörös, and Hirschberg (1996), which found a positive relationship between intangible investments and foreign direct investment. The study, however, does not directly link intellectual property rights and FDI.

Market Size

A firm's decision to invest in a foreign country is often associated with demand factors in the host country. The decision to produce abroad is linked to the potential revenues generated by sales of the product in a particular market. Therefore, the size of the market, plus its growth potential, in the target country is an important variable in determining whether foreign direct investment in a particular market will be made.

The empirical evidence with respect to the effect of market size on direct investment is convincing. In the manufacturing industry, market size was found to be a significant determinant in the decision to invest abroad (Cushman, 1985; Barrell and Pain, 1996). Since market size is often measured by the country's gross national product (GNP), the previous result may explain why most FDI is observed in developed market (Henderson and Handy, 1994).

Evidence exists that supports the market size hypothesis in the food industry as well. For example, Ning and Reed (1995) found that gross domestic product (GDP) significantly affected U.S. FDI in food and kindred products. This result was further confirmed by Gopinath, Pick, and Vasavada (1999) with respect to foreign direct investment and foreign affiliate sales. They found that a 1 percent increase in the per capita income of a country, as measured by the GDP, leads to a 0.49 percent increase in the foreign affiliate sales of processed food products. Furthermore, their study found that a 1 percent increase in the GDP of a country would lead to a 0.12 percent increase in the foreign direct investment in that country.

Other Variables

Intellectual property rights and market size are important determinants of foreign direct investment, but other variables have also been categorized as important in the decision process. Some of the variables are firm specific, whereas others are external to the firms.

One such variable that has received much attention recently is the exchange rate. The hypothesis that the exchange rate affects FDI relies on an imperfect capital market approach. If foreigners own their wealth in non-dollar denominated assets, then a depreciation of the dollar would lead to an increase in their wealth relative to economic agents in the domestic market. This, in turn, will enable foreigners to bid and acquire domestic assets more cheaply. This hypothesis was tested empirically for the manufacturing sector, among others, by Cushman (1985) and Caves (1990) who found support for a negative relationship between the dollar value and foreign direct investment.

Gopinath, Pick, and Vasavada (1998) tested the causal relationship of the exchange rate to foreign direct investment in U.S. food processing. Their results show the exchange rate to be a causal factor in the substitution between FDI and trade, and that the negative effects of the real exchange rates on processed food exports are partially offset by the increase in foreign affiliate sales of U.S. multinational corporations. Pompelli and Pick (1998) interviewed several executives in the poultry industry who identified exchange rate risk as a significant motivation for FDI.

A host of other variables have been used in empirical analyses in an attempt to explain FDI motivation. One such variable is the interest rate differential between the host and domestic markets. This variable represents the costs of capital and was used in the study by Ning and Reed who found it to significantly affect the U.S. FDI position abroad. Gopinath, Pick, and Vasavada, on the other hand, found the cost of capital in the United States to adversely affect U.S. FDI in the food-processing industry, but it did not affect foreign affiliate sales.

Another factor that affects foreign direct investment is the host country's protection policies (Lipsey and Weiss, 1984). Protection policies may discourage export and encourage the establishment of foreign affiliates. Surprisingly, several studies of FDI in U.S. manufacturing industries find that foreign trade barriers have either a weakly positive effect or no effect at all.[1] For European agriculture, Da Silva (1997) does not find a consistent effect of trade barriers on agricultural FDI. However, in their study of the U.S. food-processing industry, Gopinath, Pick, and Vasavada (1999) found protection policies, as proxied by the producer subsidy equivalents, to positively affect foreign direct investment.

Input prices associated with the production of goods in the domestic and host countries also enter the investment decision. The empirical evidence with regard to input prices is scarce, since input price data for specific industries are difficult to obtain. The study by Gopinath, Pick, and Vasavada (1999) considered three input prices in their model of the food-processing sector. Two of the variables, the wage rate and intermediate input prices, significantly affected foreign affiliate sales, while the wage rate and interest rate adversely affected foreign direct investment. Another factor influencing FDI is "cultural distance." FDI tends to go to countries with a similar language or system of laws. This is similar to the behavior of other manufacturing industries in that they agglomerate in countries where previous FDI and trade has been highest.[2] The level of a country's intellectual property protection has a similar effect as well.[3]

FUTURE RESEARCH

Several factors that may affect FDI have not been adequately studied. First among them is risk. In many countries there is an element of risk beyond usual business and currency risks. This consists of political risks such as labor unrest and government instability. Other risks include changes in tax policy or enforceability of contracts. In some developing countries there is a risk of production loss due to corruption or infrastructure disruption.

Another area for future research is the role of agricultural FDI in economic development. The role of manufacturing FDI in development has received some study, but the role of agricultural FDI has not. For many developing countries, primary agriculture claims a large share of their GDP. It may be that FDI in agricultural industries has a unique role in a country's development.

Many studies use data that are aggregated across firms and industries. Some studies, using either specific industry data or even firm-level data, demonstrate that aggregate data are not sufficient for a complete analysis of FDI. One example is Blonigen's (1999) study of the industry for specific car parts. Although disaggregated data has been used in some studies analyzing FDI in manufacturing industries, it has not been used for agricultural industries. One exception is Yilmaz's (1999) study using firm-level data from Turkey (see also Chapters 1, 2, 3, and 6).

An underused approach in studying FDI is the industry-specific study. Some aspects of FDI are unique to each industry. For example, in the wine industry, U.S. firms undertake FDI in order to secure a stable supply of wine for import to the United States (see Pompelli and Pick, 1999). In the poultry industry, however, FDI from U.S. firms is oriented toward promoting exports (see Pompelli and Pick, 1999). Significant factors unique to each industry may be missed in a cross-industry analysis.

Another area for future research is the relationship between market structure and FDI. The food-processing industry is characterized, particularly in recent years, by mergers, acquisitions, and high concentration rates. Campa, Donnenfeld, and Weber (1998) investigated this relationship for the U.S. manufacturing industries and found a negative relationship between FDI and imports in highly concentrated industries. This relationship has not yet been studied for the highly concentrated food industry.

NOTES

1. Braunerhjelm and Svensson (1996) and Hufbauer, Lakdawalla, and Malani (1994) use measures of openness and get coefficients with the expected sign, but these results are not statistically significant. Bajo-Rubio and Sosvilla-Rivero (1994) find statistically significant coefficients for Spain. For an explanation of why tests of the effect of trade barriers on FDI yield inconsistent results, see McCulloch (1993).

2. Ning and Reed (1995) and da Silva (1997) include variables for language or cultural ties. Braunerhjelm and Svensson (1996) analyze agglomeration of FDI.

3. Lee and Mansfield (1996) cite a survey of major U.S. agricultural firms showing that a significant portion find intellectual property protections in several developing countries too weak to undertake FDI.

REFERENCES

Bajo-Rubio, Oscar and Simon Sosvilla-Rivero (1994). An econometric analysis of foreign direct investment in Spain, 1964-1994. *Southern Economic Journal,* 61(1): 104-120.

Barrell, Ray and Nigel Pain (1996). An econometric analysis of U.S. foreign direct investment. *The Review of Economics and Statistics,* 78(2): 200-207.

Blonigen, Bruce A. (1999). In search of substitution between foreign production and exports. National Bureau of Economic Research working paper number 7154.

Braunerhjelm, Pontus and Roger Svensson (1996). Host country characteristics and agglomeration in foreign direct investment. *Applied Economics,* 28(7): 833-840.

Campa, J., Shabtai Donnenfeld, and Shlomo Weber (1998). Market structure and foreign direct investment. *Review of International Economics,* 6: 361-380.

Caves, R.E. (1990). Exchange rate movements and foreign direct investment in the United States. In D.R. Audretsch and M.P. Claudon (Eds.), *The internationalization of the U.S. markets* (pp. 199-229). New York: New York University Press.

Connor, J.M. (1983). Foreign investment in the U.S. food marketing system. *American Journal of Agricultural Economics,* 65: 395-404.

Cushman, D.O. (1985). Real exchange rate risk, expectations, and the level of foreign direct investment. *Review of Economics and Statistics,* 67: 297-308.

da Silva, Joao Gomes (1997). Intra European trade and investment in processed food products, 1980-1991: Changing determinants and characteristics. In Shida Rastegari Henneberry (Ed.), *Foreign Direct Investment and Processed Food Trade, Proceedings of the Conference of NCR-182* (pp. 79-100). Stillwater: Oklahoma State Department of Agricultural Economics.

Dunning, J.H. (1994). Globalization and development. Discussion paper in International Investment and Business Studies, number 187, Department of Economics, University of Reading, UK.

Gopinath, Munisamy, Daniel Pick, and Utpal Vasavada (1998). Exchange rate effects on the relationship between FDI and trade in the U.S. food processing industry. *American Journal of Agricultural Economics,* 80: 1074-1080.

Gopinath, Munisamy, Daniel Pick, and Utpal Vasavada (1999). The economics of foreign direct investment and trade in the U.S. food processing industry. *American Journal of Agricultural Economics,* 81: 442-452.

Grubaugh, Stephen G. (1987). Determinants of direct foreign investment. *The Review of Economics and Statistics,* 69: 149-152.

Henderson, Dennis R. and Charles R. Handy (1994). International dimensions of the food marketing system. In L.P. Schertz and L.M. Daft (Eds.), *Food and Agricultural Markets: The Quiet Revolution* (pp. 166-195). Washington, DC: National Planning Association.

Henderson, Dennis R., Charles R. Handy, and Steven A. Neff (1996). *Globalization of the Processed Foods Market.* Agricultural Economic Report No. 742. Washington, DC: Economic Research Service, U.S. Department of Agriculture.

Henderson, Dennis R., Peter R. Vörös, and Joseph G. Hirschberg (1996). Industrial determinants of international trade and foreign investment by food and beverage manufacturing firms. In Ian M. Sheldon and Philip C. Abbott (Eds.), *Industrial Organization and Trade in the Food Industries* (pp. 197-216). Boulder, CO: Westview Press, Inc.

Hufbauer, G.C., D. Lakdawalla, and A. Malani (1994). Determinants of foreign direct investment and its connections to trade. *UNCTAD Review,* pp. 39-51.

Lee, Jeong-Yeon and Edwin Mansfield (1996). Intellectual property protection and U.S. foreign direct investment. *The Review of Economics and Statistics,* 78(2): 181-186.

Lipsey Robert E. and Merle Yahr Weiss (1984). Foreign production and exports of individual firms. *Review of Economics and Statistics,* 66: 304-308.

McCulloch, Rachel (1993). New perspectives on foreign direct investment. A National Bureau of Economic Research project report. Chicago and London: University of Chicago Press.

Ning, Yulin and Michael R. Reed (1995). Locational determinants of the U.S. direct foreign investment in food and kindred products. *Agribusiness,* 11(1): 77-85.

Pompelli, Greg and Daniel Pick (1998). International investment motivations of U.S. broiler firms. Agricultural Marketing and Agribusiness Studies Working Paper No. SP98-17, University of Tennessee, December.

Pompelli, Greg and Daniel Pick (1999). International investment motivations of U.S. wineries. *International Food and Agribusiness Management Review,* 2(1): 47-62.

Yilmaz, Alper (1999). Host country welfare effects of foreign direct investment (FDI) and imports: An application to the processed food industry. Working paper, University of California, Davis, August.

Chapter 5

Multinational Food Corporations and Trade: The Impact of Foreign Direct Investment on Trade in the U.S. Food Industry

Andrew P. Barkley

INTRODUCTION

The level of foreign direct investment (FDI) undertaken by multinational enterprises (MNEs) is large and growing. Dunning (1998) reported that FDI flows have grown at rates more than twice as great as those of exports in the 1980s, and "by the early 1990s, the sales of foreign affiliates of multinational enterprises (MNEs) considerably exceeded those of world wide exports" (p. 43). Graham and Krugman (1995) calculated that between 1983 and 1989, world FDI flows grew at an annual compound rate of nearly 30 percent, compared to much lower annual growth rates of world trade (9 percent) and world income (8 percent). Although FDI is large and economically significant, economists have struggled to explain (1) why FDI occurs and (2) the impacts of FDI. The potential impacts on both the investing (home) nation and the country where the investment takes place (affiliate or host nation) are large. In a recent comprehensive study of the causes and consequences of FDI in the Caribbean, Barclay (2000, pp. 2-3) concluded, "For almost forty years, researchers have been grap-

---

This research was conducted while the author was Visiting Scholar at the Department of Land Economy, University of Cambridge, Cambridge, England.

pling with issues surrounding the phenomenon of the MNE and the factors that influence its behaviour in different countries."

Perhaps the most controversial feature of increasing FDI is the relationship between FDI and exports. Dunning (1993, p. 412) stated that "in today's global economy not only are trade and FDI increasingly linked with each other, but also a substantial portion of the former is undertaken by, and within, MNEs." Early studies of FDI reported that direct investments abroad were likely to reduce exports. More recent evidence, however, suggests that the levels of FDI and international trade are complementary. The actual nature of the relationship is economically important, because if FDI substitutes for trade, employment and earnings in export-competing industries may be negatively affected when firms locate abroad.

This study summarizes, supplements, and extends previous literature on the relationship between FDI and trade by empirically investigating the determinants of exports of the U.S. food-processing industry. Specifically, the impact of food-processing FDI on both agricultural and food-sector exports will be quantified using an econometric model.

The model extends previous research in several important directions. First, by specifying a single-equation regression model to quantify the empirical relationship between FDI and exports to host nations, this study provides econometric tests of the potential for positive or negative impacts of outward FDI on international trade. Second, the inclusion of separate independent variables for both real exchange rates and national income (gross domestic product [GDP]) clarifies the distinct impact of these two important determinants on U.S. agricultural and food industry exports. Third, this research investigates the impact of FDI on both food and agricultural exports, since investments in the food-processing industry are likely to affect the flow of both processed food and agricultural commodities, as agricultural goods are inputs to the production of food outputs.

The results indicate that there is an overall complementary (positive) relationship between outward FDI by U.S.-based firms and trade with host nations pooled during the time period 1989 to 1999. However, FDI in several individual host nations was negatively related to U.S. exports. Regression results demonstrated complementarity between U.S.-outward FDI and trade for nations that import agricultural goods, and the substitution of FDI and trade between the

United States and agricultural exporters (the European Union, South American nations, and Australia). The model results also suggest that the degree of openness and market size are characteristic of nations with an estimated complementary relationship between outward FDI and exports. The results, consistent with previous empirical studies, imply that a broader analytical theory of FDI is needed to explain the location and magnitude of FDI, as well as how direct investments abroad in the U.S. food-processing sector change over time.

PREVIOUS LITERATURE

Motivation for FDI

A great deal of research has been devoted to expanding our knowledge of the existence and operation of MNEs. Barclay (2000) cataloged and synthesized the previous research into the motivations for FDI. She identified four theories for FDI:

1. Monopolistic advantage
2. Oligopolistic advantage
3. Internalization
4. The "eclectic" theory

Hymer (1960) suggested that firms would engage in FDI when they have firm-specific assets, or economic advantages such as technology, finance, or distribution networks that are not possessed by their competitors in the host country. This motivation for FDI relies on the existence of imperfect markets for outputs or inputs (monopoly). Other research has emphasized strategic interactions between firms as the primary motivation for FDI (Barclay, 2000). Knickerbocker (1973) suggested that firms followed competitor firms into foreign markets as a means of protecting firm-specific assets.

The third motivation for FDI cited by Barclay is internalization, which provides a solid motivation for a firm undertaking FDI when it is more efficient (less costly) to create an internal market than to make purchases from a existing one (Coase, 1937; Williamson, 1985).

Barclay (2000) identified the fourth motivation, or the "eclectic paradigm" of FDI, with the seminal and extensive work of John H. Dunning. Dunning's "OLI theory" (ownership, location, and internalization) is the foundation for an extensive literature concerning FDI. A comprehensive survey of multinational enterprise behavior is provided in Dunning (1993). Dunning's research found that MNEs often have *ownership* benefits, such as financial resources, technology, patents, and marketing advantages that are not available in the host country. Similarly, *location* advantages can include lower wage rates, a specialized or skilled workforce, or other input advantages. Companies that locate plants within another nation may become exempt from costly trade barriers such as import tariffs or quotas. For example, as many low-income nations increase their standard of living, the opportunity to introduce products to growing markets may exist. Dunning also delineated *internalization* as a motivation for FDI. This is represented by a parent company providing a host nation with economic benefits that result from the operation of the international business through a contractual relationship such as a license, contact, franchise, or agreement.

Location of FDI

Previous literature has also informed our knowledge of where FDI is located. Minimizing the cost of production is a major determinant of the nations and localities where FDI occurs. Barclay (2000, p. 46) confirmed that "[t]here appears to be a consensus among theorists on the role that factor costs play in attracting FDI." The empirical evidence of Kumar (1994) and Woodward and Rolfe (1993) verified that a low-cost, skilled workforce can be an important determinant of FDI in many situations. Investment incentives are often provided by nations seeking to attract capital from abroad. The empirical evidence suggests, surprisingly, that financial incentive is not necessarily effective in increasing the level of FDI. Trade barriers can also result in increased FDI, as multinational firms seek to avoid the costs of import and export duties. Kumar (1994) studied the determinants of export-oriented FDI by U.S. firms. He confirmed that low-cost labor, together with location advantages such as infrastructure and services, were positively associated with FDI. Also, economies characterized by high trade barriers were able to increase the level of export-seeking

FDI. Financial incentives were not associated with changes in the level of FDI in the empirical research conducted in this area.

FDI and Trade

"The relationship between FDI and trade, though, is still a hotly debated issue" (Reed, 2001, p. 178). One of the most interesting and contentious issues in the FDI literature is the relationship between FDI and international trade. The controversy involves whether FDI is a *substitute for* or a *complement to* trade. If FDI and trade are substitutes, then firms engage in FDI as a replacement for the exportation of goods to other nations. This proposition was first put forth by Mundell (1957), with the implication that FDI results in a loss of employment in the home country due to the relocation of productive capacity to the host nation.

A second possibility is that of complementarity between FDI and exports. This could occur if home-nation products were used as inputs to the productive processes located in the host nation. In this case, the activities of a multinational firm in the home nation could be linked with host-country operations such that growth in the host-nation operation increases the demand for intermediate goods from the home nation (Gopinath, Pick, and Vasavada, 1999; Helpman and Krugman, 1985). Here, exports from the home country to the host nation could increase as resources are shipped to foreign affiliates for further processing and sales. This scenario seems intuitively likely for raw material industries, such as unprocessed food. Also, if direct investment abroad leads to economic growth, or increases in incomes, this could increase the demand for inputs and goods produced by the home nation, resulting in increased exports from the United States to the host nation.

Armed with the preceding arguments, economists have divided FDI into two types: horizontal and vertical. Horizontal FDI is defined by the case in which the same business activity is undertaken by the MNE in both the home and host countries. In this scenario, FDI and trade are likely to be substitutes, since the productive activities of the firm in the home nation are shut down and replaced by activities abroad. Vertical FDI refers to a situation in which home-country products are used as inputs to productive processes in the host nation, or outputs for final consumption in the home nation. In this situation, the relationship between FDI and trade becomes more complex and depends on the flow of inputs and outputs between home and host nations. In this scenario, vertical FDI could lead to a complementary relationship between outward FDI and trade.

Chapter 14 of Dunning (1993) thoroughly examines the previous literature concerning the relationship between FDI and trade. The pervasive theme of the survey was that it is impossible to generalize or simplify the complex relationship between the level of trade and FDI. However, numerous studies have found empirical support for a positive, complementary relationship between MNE activity and exports. Bergsten, Horst, and Moran (1978) identified a positive, significant association between exports and FDI for MNEs in the United States during the time period 1966-1977. This result was confirmed by Lipsey and Weiss, who found a complementary relationship at the aggregate level (1981) and the firm level (1984). Lipsey and Weiss (1981) found consistent evidence that the level of U.S. manufacturing affiliates was positively related to U.S. exports; they concluded, "We find no evidence that on net balance a country's production in overseas markets substitutes for its own domestic production and employment" (p. 494).

Froot and Stein (1991) examined the role of the value of the dollar on cross-border acquisitions in the United States and confirmed that the level of direct investment was associated with the real value of the dollar. The authors suggested that dollar depreciation would increase inward investment in the United States due to relative wealth changes. Klein and Rosengren (1994) augmented the theory of Froot and Stein by demonstrating that real exchange rate changes alter not only the relative wealth position of a nation but also the costs of production. Their empirical results showed that relative wealth changes completely dominated any changes in labor costs. Gopinath, Pick, and Vasavada (1998, p. 1073) reported that "a majority of studies support the negative relationship between the dollar value and FDI inflows into the United States."

McCorriston and Sheldon (1998) examined the role of the exchange rate on FDI in the U.S. food sector, based on the preceding work of Froot and Stein (1991), and Klein and Rosengren (1994). McCorriston and Sheldon's (1998, p. 1072) empirical analysis of inward FDI during the time period 1985-1995 was "consistent with the wavelike pattern of foreign acquisitions in the United States, although it is apparent that the exchange rate is not the only relevant variable." Other studies of the impact of exchange rates on exports, outward FDI, and foreign affiliate sales include Hooper and Kohlagren (1978), who studied exports, and Stevens (1998), who explored foreign affiliate sales.

Gopinath, Pick, and Vasavada (1998) tested the hypothesis that FDI and exports were substitutes in the U.S. food-processing industry for ten nations during 1982-1995. The authors found that "real exchange rates have a positive effect on outward FDI and foreign sales by U.S. majority-owned multinational food companies" (p. 1077). Gopinath, Pick, and Vasavada (1998) suggested that a complementary relationship between exports and FDI might exist for three reasons. First, foreign opportunities are dynamic. Second, final product exports from the home country may decrease but could be offset by exports of intermediate products. Third, if technological differences exist, it may be possible for FDI and trade to simultaneously increase.

Blomstrom, Lipsey, and Kulchyck (1988), using data from the United States and Sweden, concluded that the relationship between foreign production and exports was somewhere between neutrality and complementarity. Pearce (1990) confirmed this relationship for 1982 data on exports and foreign production. Agerwal (1987) reached a similar conclusion for data from India. Not only does most of the evidence suggest a complementary relationship between trade and FDI, but Dunning (1993) found that MNEs tend to concentrate in trade-intensive sectors. In the next section, direct investment in the manufactured food industry will be explored.

FDI in the Food Industry

Chapter 11 in Reed (2001) provides an outstanding introduction to the magnitude of FDI in the food industry, as well as the major issues surrounding it. Reed (2001) reported that "[a]ctivities of U.S. food MNEs are huge and growing" (p. 179). Rama (1998a) examined this growth by delineating the internal forces associated with long-run global growth in large multinationals in the food and drink processing industry during the 1977-1988 time period. Her empirical results showed that increasing multinationality was associated with diminishing firm growth rates, implying that smaller, newer firms may develop faster than large, established multinationals. Further research by Rama indicated that firms in the food and drink industry should consider exporting (1998b) and developing new technology (1996) to avoid productive inertia.

Processed food exports from the United States have increased in recent years, primarily due to lower trade barriers and increased living standards throughout the world (Reed, 2000). Trade in processed food was thoroughly investigated by Henderson, Handy, and Neff (1996), who surveyed FDI and trade in the food sector of the United States. Handy and Henderson (1994) found that most FDI undertaken by the U.S. food industry is horizontal in nature, indicating that the type of business activity is the same for both the parent company and the affiliate. Several recent studies have investigated the relationship between FDI and exports in the food-processing industry. Ning and Reed (1995) found that affiliate sales and exports are substitutes in the food sector, whereas Overend, Connor, and Salin (1995) and Munirathinam, Reed, and Marchant (1999) found a complementary relationship in some cases. Reed and Marchant (1992) found that the food-processing industry's international competitiveness was falling behind other manufacturing industries in the United States.

AN EMPIRICAL MODEL OF THE IMPACT OF FDI ON INTERNATIONAL TRADE

An empirical model was developed to quantify the relationship between outbound FDI on exports in the food processing and agricultural sectors of the United States. Previous literature focused on the impact of exchange rates on exports and FDI separately. The impact of FDI on exports is measured directly, as in Equation 5.1, where i = nation i, and t = time period t.

$$\text{EXPORTS}_{it} = f(\text{FDI}_{it}, \text{REAL EXCHANGE RATE}_{it}, \text{GDP}_{it}, \text{GDP}_{usat}) \tag{5.1}$$

The model diverges from previous literature in several important ways. First, the regression model directly measures the impact of FDI on exports by placing both in a single equation, with exports as the dependent variable. This specification quantifies the proposition that direct investment impacts the level of exports, in either a positive (complementary) or negative (substitution) direction.

Second, Gopinath, Pick, and Vasavada (1998) suggested that a real, constant-dollar exchange rate is superior to a nominal exchange rate in accounting for inflation in both nations. Therefore, a real exchange rate is included in the regression model here. Third, the specification here diverges from previous studies by incorporating national income (GDP) as a separate independent (right-hand side) variable, rather than measuring FDI and exports as a percentage of GDP. The specification in Equation 5.1 is more general and allows for the direct measurement of how changes in income influence export levels and how real exchange rates affect the level of exports, rather than the percentage of GDP accounted for by exports. This specification isolates the impact of the real exchange rate on exports by holding constant the real GDP of both the United States and the importing (host) nation.

Equation 5.1 is estimated using ordinary least squares (OLS) and is subject to the possibility of collinearity between the two GDP measures, GDP_{it} and GDP_{usa}, due to possible correlation in income across nations. To account for this possibility, the conceptual model in Equation 5.1 is specified as the regression model in Equation 5.2, where GDPRATIO is defined as GDP_{it} divided by GDP_{usa}, and e is the error term.

$$EXPORTS_{it} = a + b_1 FDI_{it} + b_2 \text{ REAL EXCHANGE RATE}_{it} + b_3 GDPRATIO_{it} + e_{it} \tag{5.2}$$

The model includes what are considered to be the most important determinants of exports. However, due to the high level of aggregation, the model overlooks many nation-specific factors such as trade barriers, transactions costs, or institutional differences that could influence international trade between nations. These factors are accounted for by including a qualitative (0-1) variable for each nation as an independent variable (intercept shifter) in the regression model of Equation 5.2 ($a = a_i$), resulting in the fixed coefficients (model two). A third model tests for the possibility of variable FDI coefficients (slope shifters) by estimating the coefficient on FDI (b_1) separately for each nation (model three).

Previous studies have emphasized the possibility of a complementary relationship between FDI and exports in the food-processing industry resulting from vertical trade relationships. Specifically, outbound FDI in U.S. food processing could be enhanced by increased exports of agricultural products as food processing in the host nation increases. To account for this possibility, the impact of FDI by U.S.-based food-processing firms on exports of both manufactured food exports and agricultural exports is quantified using the regression model specified previously. Together, these regressions provide more complete information about the impact of FDI on the flow of goods across national boundaries than was previously available.

DATA

The regression model is tested using FDI data in the U.S. food-processing sector provided by the U.S. Department of Commerce, Bureau of Economic Analysis (2002). The food-processing industry is classified by the Standard Industrial Classification (SIC) system as SIC20, the largest manufacturing industry in the United States, representing approximately 14 percent of the manufactured output of the United States (Reed, 2000). The food-processing industry is composed of nine categories, including meat products (SIC201), dairy products (SIC202), preserved fruits and vegetables (SIC203), grain mill products (SIC204), bakery products (SIC205), sugar and confectionery products (SIC206), fats and oils (SIC207), beverages (SIC208), and miscellaneous products (SIC209). Complete data are available for twenty-three nations over the time period 1989 to 1999. Export data are also available for this time period (U.S. International Trade Commission, 2002) for both the food-processing sector (SIC20) and agricultural goods. Exchange-rate data are from the International Monetary Fund's (IMF) World Economic Outlook Databank (2002) and are deflated using GDP deflators from the same source. Gross domestic product data are from the IMF and are in real, constant-dollar (1996) values for both the United States and the twenty-three importing nations.

Foreign direct investment by U.S.-based food-processing firms is reported in Table 5.1. Manufactured-food FDI grew from $14.9 billion in 1989 to over $35 billion in 1999, demonstrating the large and increasing economic importance of U.S.-based food-processing firms. Growth rates of FDI were large in all of the world regions over the eleven-year period. The comparison of processed-food FDI (SIC20) relative to total FDI for all U.S. industries is shown in Table 5.2. These data include manufactured food but not retail food outlets

TABLE 5.1. U.S. direct investment position in food and kindred products (SIC 20), in $ million, 1999.

Region	1989	1990	1991	1992	1993	1994	1995	1996	1997	1998	1999
Latin America (including Mexico) and other Western Hemisphere	2,974	3,611	4,128	4,651	6,231	6,592	7,706	9,198	9,390	10,017	9,710
Europe	7,030	9,475	9,361	11,895	14,460	12,462	14,695	15,012	15,082	16,574	16,094
South America	1,962	2,020	2,205	2,672	3,171	3,398	4,314	5,133	4,409	4,786	4,154
Central America	960	1,536	1,852	1,926	3,007	3,152	3,348	4,023	4,938	5,190	5,406
Asia and Pacific	2,227	2,541	3,077	3,674	3,684	3,249	3,289	3,354	3,794	3,468	3,647
All countries	14,943	18,834	20,015	24,089	28,774	27,126	30,826	32,467	33,643	35,800	35,151

Source: U.S. Department of Commerce, Bureau of Economic Analysis, 2002.

175

TABLE 5.2. U.S. direct investment position, manufactured food products and all industries, 1999.

Industry	1989	1990	1991	1992	1993	1994	1995	1996	1997	1998	1999
All (in $ billions)	480	521	546	572	628	668	746	832	894	1,015	1,131
Food (in $ millions)	14,943	18,834	20,015	24,089	28,774	27,126	30,826	32,467	33,643	35,800	35,151
Food:all ratio	0.031	0.036	0.037	0.042	0.045	0.041	0.041	0.037	0.035	0.035	0.031

Source: U.S. Department of Commerce, Bureau of Economic Analysis, 2002.

(U.S. Department of Commerce, 2002). Interestingly, FDI in food products and total FDI both grew at the same rate of approximately 135 percent during the period under investigation, 1989 to 1999, an average annual increase of 3.71 percent (see Table 5.2). The relationship between manufactured-food FDI and exports is illuminated in Table 5.3, which reports a decline in agricultural exports simultaneous with an increase in food exports from 1989 to 1999. The ratio of agricultural to food exports declined from 1.65 in 1989 to 0.86 in 1999 (see Table 5.3), demonstrating changes in the structure of U.S. food exports. Interestingly, the average annual decrease in agricultural exports is equal to 1.42 percent, and the average annual increase in food industry exports was 1.40 percent over the eleven-year period 1989 to 1999.

The data presented in Tables 5.2 and 5.3 suggest that the large increase in FDI in the food-processing sector may be complementary with food exports but substitute for or crowd out agricultural exports. The regression analyses presented here test for this possibility by quantifying the marginal effect of FDI on processed-food and agricultural exports, holding income and exchange rates constant. The data employed in the econometric analysis are summarized in Table 5.4. The total number of observations is 250, three less than a complete set of twenty-three nations and eleven years, because exchange rate data were unavailable for Brazil during the years 1989 to 1991. The Brazil data for the remaining years, 1992-1999, are included in the regressions. Other nations without complete data available were excluded from the analysis. Brazil was retained due to the relatively large flow of FDI to that nation (see Chapter 9), in spite of the three missing observations, to make the estimated coefficients as inclusive as possible. The inclusion of Brazil leads to greater degrees of freedom and reduces bias in the estimated coefficients since it is a major trading partner of food and agricultural products from the United States.

REGRESSION RESULTS

The results of the regression models for agricultural exports are reported in Table 5.5. The reported standard errors were corrected for heteroskedasticity using White's (1980) consistent estimator. Model

TABLE 5.3. Agriculture and food industry (SIC 20) exports, in $million, 1999.

Industry	1989	1990	1991	1992	1993	1994	1995	1996	1997	1998	1999
Ag	31,883	29,099	26,500	27,499	25,671	26,048	33,202	35,720	30,222	25,630	22,982
Food	19,323	19,747	20,690	22,856	23,198	25,571	28,164	28,850	29,882	28,279	26,648
Ag:food ratio	1.65	1.47	1.28	1.20	1.11	1.02	1.18	1.24	1.01	0.91	0.86

Source: U.S. International Trade Commission, 2002.

TABLE 5.4. Summary statistics of variables included in food industry export regressions.[a]

Variable	Mean	Standard deviation	Minimum	Maximum
U.S. ag exports (1996 in $ billions)	797.263	1189.97	14.507	6,211.99
U.S. food mfg exports (1996, in $ billions)	735.916	1228.56	11.828	5,732.82
GDP ratio (nation i/U.S.)	0.089	0.127	0.001	0.71
Real exchange rate (currency i/$)	279.943	544.193	0.544	2,618.46
U.S. food FDI (1996, in $ billions)	1.005	1.174	0.001	4.83
Argentina	0.044	0.206	0	1.00
Australia	0.044	0.206	0	1.00
Belgium	0.044	0.206	0	1.00
Brazil[b]	0.032	0.176	0	1.00
Canada	0.044	0.206	0	1.00
Chile	0.044	0.206	0	1.00
China	0.044	0.206	0	1.00
Colombia	0.044	0.206	0	1.00
Costa Rica	0.044	0.206	0	1.00
Finland	0.044	0.206	0	1.00
France	0.044	0.206	0	1.00
Germany	0.044	0.206	0	1.00
Guatemala	0.044	0.206	0	1.00
Italy	0.044	0.206	0	1.00
Japan	0.044	0.206	0	1.00
Korea	0.044	0.206	0	1.00
Mexico	0.044	0.206	0	1.00
Netherlands	0.044	0.206	0	1.00
Philippines	0.044	0.206	0	1.00
Spain	0.044	0.206	0	1.00
Thailand	0.044	0.206	0	1.00
United Kingdom	0.044	0.206	0	1.00
Venezuela	0.044	0.206	0	1.00

Source: Agricultural and food manufacturing exports: U.S. International Trade Commission; FDI and GDP: U.S. Department of Commerce, BEA; Exchange rate: International Monetary Fund (IMF) World Economic Outlook Databank.
[a]The number of observations equals 250: twenty-three nations for the eleven-year period 1989-1999, with three missing observations for Brazil, 1989-1991.
[b]Exchange rate data were unavailable for Brazil during the period 1989-1991. These three years were therefore omitted from the regression analyses presented in Tables 5.5 and 5.6.

one, which includes the three major explanatory variables (FDI, the real exchange rate, and the GDP ratio) explained approximately 51 percent of the variation in U.S. agricultural exports. The estimated coefficient on FDI is positive and statistically significant, indicating that the overall relationship between processed-food FDI and agricultural exports is complementary. The real exchange rate was not statistically significant, whereas the GDP ratio was large and highly statistically significant. When GDP for the United States and the importing nation were included as separate variables in the regression, the estimation was subject to multicollinearity, with high condition indices (Belsley, Kuh, and Welch, 1980). The results of Table 5.5 provide empirical evidence that an increase in national income in the importing nation relative to the United States (as measured by the GDP ratio) results in an increase in agricultural exports from the United States. The influence of income is much larger than that of FDI, indicating that FDI does not alter agricultural trade flows to the degree that changes in the standard of living do. Real exchange rates did not influence agricultural exports in model one. This result is likely due to misspecification (omitted variable bias), since national differences in trade levels are not taken into account in model one.

The results of the fixed coefficients regression (model two) demonstrate that over 95 percent of the variation in agricultural exports was accounted for by the variables included in model one, together with the nation-specific intercepts reported in Table 5.5. Finland was selected as the default nation, as it had the smallest average level of agricultural exports from the United States during the time period covered in this analysis. In twenty of the twenty-two included nations, the difference in agricultural export levels was statistically significant relative to the default level of exports in Finland.

The fixed-effect model includes an "intercept shifter" for each nation to test for the possibility of statistical differences in the intercept of the regression, which in this case captures the quantitative value of the dependent variable, agricultural exports, in the regression model. Thus the twenty statistically significant national qualitative variables reported in model two (see Table 5.5) quantify differences in the dollar value of agricultural exports from the United States to the nations listed in Table 5.5. The relatively large, positive estimated coefficients for Japan, Canada, Mexico, and Korea captured the high value

TABLE 5.5. Agricultural export regression results.[a]

Variable	Model one Parameter estimate	t-statistic	Model two fixed coefficients Parameter estimate	t-statistic	Model three variable coefficients Parameter estimate	t-statistic
Intercept	78.635	1.524	−23.467	−0.895	224.491	2.537[c]
FDI	159.330	2.132[d]	149.889	3.347[c]	–	–
Real ex rate	0.001	0.017	−0.101	−1.578[f]	−0.019	−0.249
GDP ratio	6243.987	8.824[c]	2655.078	1.752[e]	5501.401	4.304[c]
Argentina	–		−115.983	−2.683[c]	−419.253	−4.342[c]
Australia	–		−189.082	−2.715[c]	−342.689	−5.285[c]
Belgium	–		169.470	4.024[c]	−89.592	−0.900
Brazil	–		−314.830	−1.989[e]	−251.277	−4.130[c]
Canada	–		1810.923	7.110[c]	486.112	18.472[c]
Chile	–		107.855	2.815[d]	−1587.031	−2.322[d]
China	–		753.304	3.834[c]	1069.422	0.834
Colombia	–		342.085	3.038[c]	−62.529	−0.141
Costa Rica	–		157.759	5.091[c]	−1203.920	−1.275
Finland[b]	–		–	–	−40381.000	−3.482[c]
France	–		−563.929	−2.016[e]	−412.263	−4.558[c]
Germany	–		−268.969	−0.654	−707.068	−3.473[c]
Guatemala	–		101.192	4.008[c]	−2302.137	−1.565
Italy	–		78.552	0.310	−672.720	−2.890[c]
Japan	–		3584.999	3.896[c]	2448.767	12.359[c]
Korea	–		1494.644	7.554[c]	2531.417	42.741[c]
Mexico	–		1747.298	10.512[c]	585.281	13.958[c]
Netherlands	–		766.239	9.470[c]	420.381	3.199[c]
Philippines	–		331.410	10.542[c]	337.562	4.115[c]
Spain	–		319.450	2.705[c]	23.781	0.370
Thailand	–		250.243	12.794[c]	−315.272	−0.372
United Kingdom	–		−448.165	−1.746[e]	−199.725	−3.768[c]
Venezuela	–		279.599	6.401[c]	36.652	0.199
R-square		0.512		0.957		0.919
Adj. R-square		0.506		0.952		0.910

TABLE 5.5 *(continued)*

Variable	Model one Parameter estimate	Model one t-statistic	Model two fixed coefficients Parameter estimate	Model two fixed coefficients t-statistic	Model three variable coefficients Parameter estimate	Model three variable coefficients t-statistic
F value		85.915[c]		199.519[c]		101.672[c]
Root MSE		836.622		260.096		357.047
Observations		250		250		250

Note: t-statistics are heteroskedasticity consistent (White); diagnostics indicated that degrading collinearity is not present (Belsley, Kuh, and Welch).
[a]dependent variable: agricultural exports (SIC 20); dependent variable mean = 797.263.
[b]Finland is the default nation, omitted from regression model two.
[c]$p = 0.01$.
[d]$p = 0.05$.
[e]$p = 0.10$.
[f]$p = 0.15$.

of agricultural goods flowing from the United States to these nations relative to the default nation (Finland).

The estimated coefficient for the real exchange rate variable equaled –0.101, was statistically significant at the 15 percent level, and was of the anticipated sign. The addition of the national qualitative variables reduced the misspecification error of model one and provided a marginally statistically significant estimated coefficient for the real exchange rate. This result is consistent with the findings of Gopinath, Pick, and Vasavada (1998, p. 1078): "Real exchange rates have been shown to be a causal factor in this apparent substitution between FDI and trade."

The estimated coefficients for FDI in each nation of the variable coefficients model (model three) indicate that although the overall relationship between food-processing FDI and agricultural exports is complementary, the sign of the relationship differs across nations. In the variable coefficients model, the national variables represent "slope shifters," or explanatory variables that allow for statistical differences in the slope of the relationship between the independent variable FDI and the dependent variable, agricultural exports. The model three results, therefore, capture differences across nations in the empirical relationship between outward FDI from the United States and agricultural exports from the United States to host nations.

The results are fascinating but intricate: Six nations had a statistically significant, positive relation between food FDI and agricultural exports (Canada, Japan, Korea, Mexico, the Netherlands, and the Philippines). However, nine nations were characterized by a negative, or substitute, relationship between processed-food FDI and agricultural exports: Argentina, Australia, Brazil, Chile, Finland, France, Germany, Italy, and the United Kingdom. This result reflects the complexity of foreign investment decisions and the variety of outcomes from overseas investments in food processing. The circumstances that lead to FDI are complex and dynamic, and the resulting changes in international trade can be intricate and at times divergent. Therefore, it should come as no surprise that the impact of the flow of goods between nations is not uniform and consistent across all nations in all circumstances.

Given the six nations that have positive estimated coefficients in model three (Canada, Japan, Korea, Mexico, the Netherlands, and the Philippines), the level of openness to trade and low transportation costs appear to be associated with higher levels of agricultural exports. Economic growth appears to be associated with a complementary relationship, since these six nations were characterized by relatively high GDP growth rates during the period under investigation, 1989 to 1999. Negative estimated coefficients occurred in high-income nations in Europe (Finland, France, Germany, Italy, and the United Kingdom), three nations in South America (Argentina, Brazil, and Chile), and Australia. This interesting result reflects food-industry investment abroad in nations that export agricultural commodities. The general pattern that emerges from the regression model results is a complementary relationship between outward FDI and trade with importers of U.S. agricultural goods, and a negative relationship for those nations that export agricultural commodities. Interestingly, a variety of income levels are represented by nations characterized by both positive and negative estimated relationships between outward FDI and agricultural exports.

Regression results for the impact of food-processing FDI on food-manufacturing exports are reported in Table 5.6. The qualitative results for food-manufacturing exports are identical to those for agricultural-good exports. Relative income changes, as captured by the

TABLE 5.6. Food industry manufacturing export regression results.[a]

Variable	Model one		Model two fixed coefficients		Model three variable coefficients	
	Parameter estimate	t-statistic	Parameter estimate	t-statistic	Parameter estimate	t-statistic
Intercept	−102.785	−1.903[e]	16.174	0.712	77.761	1.070
FDI	351.621	3.900[c]	236.781	3.852[c]	−	−
Real ex rate	0.002	0.022	−0.062	−2.465[d]	0.002	0.038
GDP ratio	5425.401	7.379[c]	1945.070	1.588[f]	4130.980	3.535[c]
Argentina	−		−181.560	−3.809[c]	−164.090	−2.451[c]
Australia	−		−88.867	−1.216	6.899	0.152
Belgium	−		2.563	0.058	49.310	0.703
Brazil	−		−539.296	−3.676[c]	−153.407	−3.195[c]
Canada	−		2705.909	6.888[c]	878.914	29.513[c]
Chile	−		33.351	1.485	−460.728	−0.904
China	−		75.215	0.689	−410.144	−0.660
Colombia	−		142.121	2.708[c]	86.796	0.312
Costa Rica	−		13.781	0.567	−548.703	−0.716
Finland[b]	−		−	−	−10994.000	−1.278
France	−		−539.403	−2.230[d]	−216.993	−2.983[c]
Germany	−		−372.987	−1.160	−453.315	−2.481[c]
Guatemala	−		81.751	3.392[c]	668.618	0.563
Italy	−		−160.205	−0.865	−499.092	−2.663[d]
Japan	−		3458.451	5.270[c]	3221.122	4.037[c]
Korea	−		1099.340	13.522[c]	2389.584	4.766[c]
Mexico	−		1306.165	6.918[c]	572.654	16.013[c]
Netherlands	−		468.295	5.720[c]	435.917	3.905[c]
Philippines	−		141.415	3.972[c]	431.755	2.079[d]
Spain	−		−156.140	−1.529	−49.427	−1.056
Thailand	−		62.821	3.700[c]	−163.670	−0.257
United Kingdom	−		−416.483	−1.752[e]	−14.821	−0.328
Venezuela	−		109.100	3.079[c]	148.127	1.015
R-square		0.505		0.964		0.954
Adj. R-square		0.499		0.960		0.949

TABLE 5.6 *(continued)*

Variable	Model one Parameter estimate	Model one t-statistic	Model two fixed coefficients Parameter estimate	Model two fixed coefficients t-statistic	Model three variable coefficients Parameter estimate	Model three variable coefficients t-statistic
F value		83.685[c]		238.711[c]		187.281[c]
Root MSE		869.547		246.370		276.777
Observations		250		250		250

Note: t-statistics are heteroskedasticity consistent (White); diagnostics indicated that degrading collinearity is not present (Belsley, Kuh, and Welch).
[a]dependent variable: food industry manufacturing exports (SIC 20); dependent variable mean = 735.916.
[b]Finland is the default nation, omitted from regression model two.
[c]$p = 0.01$.
[d]$p = 0.05$.
[e]$p = 0.10$.
[f]$p = 0.15$.

GDP ratio, were the most statistically significant determinant of food-manufacturing exports. The regression results indicate an over-all complementary relationship between FDI and manufactured-food exports from the United States for the included host nations (Table 5.6). As in the agricultural export model one regression reported in Table 5.5, the real exchange rate was not statistically significant, probably due to misspecification of model one since the model omitted the national intercept shifters.

Model two demonstrates that national differences account for much of the variation in manufactured-food exports. Most interesting, however, are the different signs on the variable FDI coefficients (slope shifters) reported in model three (Table 5.6). Six nations had complementary relationships between FDI and exports (Canada, Japan, Korea, Mexico, the Netherlands, and the Philippines). Note that these are the same six nations that had positive estimated coefficients in the agricultural export model reported in Table 5.5. Five nations were characterized by an inverse relationship between FDI and exports: Argentina, Brazil, France, Germany, and Italy. The real exchange rate was negative and strongly statistically significant in the fixed effects model for manufactured food (model two, Table 5.6).

The negative estimated coefficient shows that dollar appreciation leads to lower levels of exports, because the price of dollars in terms of other currencies has risen. The real exchange rate, however, was not found to be statistically significant in the variable coefficients model (model three, Table 5.6), indicating that the national FDI coefficients, when included in the regression model, had a greater influence on U.S. food exports than changes in the real exchange rate. Financial crises in Mexico (1994) and East Asia (1997) occurred during the time period under investigation. These crises may have influenced the estimated coefficients for real exchange rates, since five of the included nations (Argentina, Brazil, Korea, Mexico, and Thailand) were affected by the crises. Lipsey (2001) concluded that direct investment abroad declined in the short term but remained remarkably stable during these financial crises.

The exchange rate results reported here confirm and extend previous research by McCorriston and Sheldon (1998) and Gopinath, Pick, and Vasavada (1999) by including income levels as an independent variable separate from exchange rates. This allows for the possibility of separate estimated effects of income changes and real exchange rate changes. The regression results were sensitive to model specification: the real exchange rate was statistically significant in three of the six regression models, and the GDP ratio was statistically significant in all six regressions reported in Tables 5.5 and 5.6. Further examination of the impact of real exchange rates would be a fruitful area of future research. Previous research has emphasized the possibility of including measures of exchange rate volatility and/or expected exchange rates in empirical models (Cushman, 1985).

Similar to the agricultural export models, the manufactured food models results reflect the complexity of FDI impacts. Empirical results reported in Tables 5.5 and 5.6 indicate that FDI is positively related to exports of both agricultural and food products from the United States. Complementarity between FDI and food exports occurred between the United States and importers of agricultural commodities. A negative (substitute) relationship between FDI and food exports was estimated for nations that export agricultural commodities (nations from the European Union and South America).

IMPLICATIONS AND CONCLUSIONS

This chapter has reviewed the previous literature on the relationship between FDI and exports in the food-processing industry. Earlier research emphasized a number of reasons why direct investment abroad occurs and that the impacts of FDI are varied and complex. An econometric model was specified to identify and quantify the determinants of agricultural and food industry exports from the United States. The six regressions reported here provide empirical evidence that relative income growth, as measured by the GDP ratio, was associated with higher levels of agricultural and food exports from the United States. In three of the six regressions, the real exchange rate was shown to influence the level of processed food exports in the expected direction, confirming the outcome of previous research on the influence of exchange rate movements on international trade flows.

Previous research has tried to identify whether FDI results in a decrease in exports, due to substitution of investment for exports, or an increase in exports, due to the possible complementarity of foreign direct investment and exports. The major pattern that emerged was the estimated complementarity between U.S. FDI and trade for host nations that import agricultural goods, and the substitution of FDI and trade between the U.S. and agricultural exporters (the European Union, South American nations, and Australia). This study has expanded our knowledge about the empirical relationship between outward FDI and exports in food and agricultural sectors. There is still much left to learn, however. Gopinath, Pick, and Vasavada (1998) called for examination of the relationship between FDI and trade at a disaggregate level, to include such possible determinants of outward FDI as wage rates, tax rates, market size, transport costs, environmental policy, measures of openness, and risk. The results of this study indicate that these factors are likely to explain the level and variability of FDI. Specifically, the regression results suggest that the degree of openness and market size are characteristic of nations with an estimated complementary relation between outward FDI and exports. Future work that extends our knowledge by identifying the determinants of FDI and clarifying the relationship between FDI and international trade is likely to yield meaningful and productive results.

REFERENCES

Agarwal, J.P. (1987). The strategic challenge of the evolving global economy. *Business Horizons* 30(4): 38-44.

Barclay, Lou Anne A. (2000). *Foreign direct investment in emerging economies.* London: Routledge.

Belsley, D., E. Kuh, and R.E. Welsch (1980). *Regression diagnostics.* New York: John Wiley and Sons.

Bergsten, C.F., T. Horst, and T.H. Moran (1978). *American multinationals and American interests.* Washington, DC: The Brookings Institute.

Blomstrom, M., R.E. Lipsey, and K. Kulchyck (1988). U.S. and Swedish direct investment and exports. In R.E. Baldwin (Ed.), *Trade policy issues and empirical analysis* (pp. 259-297). Chicago: University of Chicago Press.

Coase, R.H. (1937). The nature of the firm. *Economica* 4: 386-405.

Cushman, D.D. (1985). Real exchange rate risk, expectations, and the level of direct investment. *Review of Economics and Statistics* 67: 297-308.

Dunning, John H. (1993). *Multinational enterprises and the global economy.* Wokingham, England: Addison-Wesley Publishing Company.

Dunning, John H. (1998). The changing geography of foreign direct investment: Explanations and implications. In Nagesh Kumar (Ed.), *Globalization, foreign direct investment, and technology transfers* (pp. 43-89). London: Routledge.

Froot, K.A. and J.C. Stein (1991). Exchange rates and foreign direct investment: An imperfect capital markets approach. *Quarterly Journal of Economics* 106: 1191-1217.

Gopinath, Munisamy, Daniel Pick, and Utpal Vasavada (1998). Exchange rate effects on the relationship between FDI and trade in the U.S. food processing industry. *American Journal of Agricultural Economics* 80(5): 1073-1079.

Gopinath, Munisamy, Daniel Pick, and Utpal Vasavada (1999). The economics of foreign direct investment and trade with an application to the U.S. food processing industry. *American Journal of Agricultural Economics* 81: 442-452.

Graham, E.M. and P.R. Krugman. (1995). *Foreign direct investment in the United States.* Washington, DC: Institute for International Economics.

Handy, Charles and Dennis Henderson (1994). Assessing the role of foreign direct investment in the food manufacturing industry. In M.E. Bredahl, P.C. Abbott, and M.R. Reed (Eds.), *Competitiveness in international food markets* (pp. 203-230). Boulder, CO: Westview Press.

Helpman, E. and P.R. Krugman (1985). *Market structure and foreign trade: Increasing returns, imperfect competition, and the international economy.* Cambridge, MA: MIT Press.

Henderson, Dennis, Charles Handy, and Steven Neff (1996). Globalization of the processed food market. Agricultural Economic Report Number 742. Washington, DC: Economic Research Service, U.S. Department of Agriculture.

Hooper, P. and S.W. Kohlagren (1978). The effects of exchange rate risk and uncertainty on the prices and volume of trade. *Journal of International Economics* 8: 483-511.

Hymer, S.H. (1960). *The international operations of national firms: A study of direct investment* [doctoral thesis]. Cambridge, MA: MIT Press.

International Monetary Fund (IMF) (2002). The world economic outlook (WEO) databank. <www.imf.org>.

Klein, M.W. and E.S. Rosengren (1994). The real exchange rate and foreign direct investment in the United States: Relative wealth vs. relative wage effects. *Journal of International Economics* 36: 373-389.

Knickerbocker, F.T. (1973). *Oligopolistic reaction and the multinational enterprise*. Cambridge, MA: Harvard University Press.

Kumar, N. (1994). Determinants of export orientation of foreign production by U.S. multinationals: An inter-country analysis. *Journal of International Business Studies* 25(1): 144-156.

Lipsey, R.E. (2001). Foreign direct investment in three financial crises. National Bureau of Economic Research (NBER) working paper no. w8084. <http://papers.nber.org/papers/w8084>.

Lipsey, R.E. and M.Y. Weiss (1981). Foreign production and exports in manufacturing industries. *Review of Economics and Statistics* 63: 488-494.

Lipsey, R.E. and M.Y. Weiss (1984). Foreign production and exports of individual firms. *Review of Economics and Statistics* 66: 304-308.

McCorriston, Steve and Ian M. Sheldon (1998). Cross-border acquisitions and foreign direct investment in the U.S. food industry. *American Journal of Agricultural Economics* 80(5): 1066-1072.

Mundell, Robert (1957). International trade and factor mobility. *American Economic Review* 67: 321-335.

Munirathinam, Ravi, Michael Reed, and Mary Marchant (1999). Effects of the Canada/U.S. free trade agreement on U.S. agricultural exports. *International Food and Agribusiness Management Review* 1(3): 403-415.

Ning, Yulin and Michael Reed (1995). Locational determinants of U.S. direct foreign investment in food and kindred products. *Agribusiness: An International Journal* 11(1): 77-85.

Overend, C., John Connor, and Victoria Salin (1995). Foreign direct investment and U.S. exports of processed foods: Complements or substitutes? Paper presented at the Organization and Performance of World Food Systems (NCR-182) Spring Conference, Arlington, VA, March 9-10.

Pearce, R.D. (1990). Overseas production and exporting performance: Some further investigations. University of Reading Discussion Papers in International Investment and Business Studies, Series B, No. 135.

Rama, Ruth (1996). Empirical study on sources of innovation on international food and beverage industry. *Agribusiness* 12(2): 123-134.

Rama, Ruth. (1998a). Growth in food and drink multinationals, 1977-94: An empirical investigation. *Journal of International Food and Agribusiness Marketing* 10(1): 31-52.

Rama, Ruth. (1998b). Productive inertia and technological flows in food and drink processing. *International Journal of Technology Management* 16(7): 689-694.

Reed, Michael R. (2000). The food processing industry. In Dale Colyer, P. Lynn Kennedy, William A. Amponsah, Stanley M. Fletcher, and Curtis M. Jolly (Eds.), *Competition in agriculture: The United States in the world market* (pp. 251-268). Binghamton, NY: The Haworth Press, Inc.

Reed, Michael R. (2001). *International trade in agricultural products.* Upper Saddle River, NJ: Prentice-Hall.

Reed, Michael R. and Mary Marchant (1992). The global competitiveness of the U.S. food processing industry. *Northeastern Journal of Agricultural and Resource Economics* 22(1): 61-70.

Stevens, G.V.G. (1998). Exchange rates and foreign direct investment: A note. *Journal of Policy Modeling* 20: 393-401.

United States Department of Commerce, Bureau of Economic Analysis (BEA) (2002). U.S. Direct Investment Abroad. Balance of Payments and Direct Investment Position Data (Table 17, U.S. Direct Investment Position Abroad, Food and Kindred Products SIC 20).

United States International Trade Commission (USITC) (2002). USITC tariff and trade data. <http://dataweb.usitc.gov>.

White, H. (1980). A heteroskedasticity-consistent covariance matrix estimator and a direct test for heteroskedasticity. *Econometrica* 48: 817-838.

Williamson, O.E. (1985). *The economic institutions of capitalism.* New York: Free Press.

Woodward, D. and R. Rolfe (1993). The location of export oriented foreign direct investment in the Caribbean basin. *Journal of International Business Studies* 24(1):121-144.

Chapter 6

External versus Internal Markets of the Multinational Enterprise: Intrafirm Trade in French Multinational Agribusiness

Emmanuelle Chevassus-Lozza
Jacques Gallezot
Danielle Galliano

In the traditional trade theory of comparative advantages, capital and labor are considered to be national assets, assumed not to cross national borders. The fundamental difference between national systems is related to the differences between their factor endowment, and trade is simply explained by factorial differences between nations. Though the firm is the main actor in foreign trade, it is hardly mentioned in orthodox international economics, and it is also absent from the macroeconomic approach. It is only thanks to the new international economics and the developments in industrial economics that the role of the firm has been recognized and taken into account. Yet companies are at the center of the current debate on the internationalization of economies, and any analysis of foreign trade must take into account the efficiency of firms organized into multinational groups, particularly those that have affiliates abroad. As a result of the important development of foreign direct investments (FDI), this type of organization has generated, at the international level, internal flows of capital, technology, jobs, services, and goods. Thus the existence of intrafirm flows accentuates the necessity to take into account the internal organization of multinational enterprises (MNEs) and raises the fundamental question of their ability to bypass both the market and the borders of the nation.

But if intrafirm trade points out the fact that internal organization does matter, unfortunately data on this type of trade are often not available for European countries. The only systematic information is that produced by the U.S. Department of Commerce and the Japanese Ministry of Foreign Trade and Industry (MITI) since the 1970s. Where the European community (EC) is concerned, information is very limited and given at a global level. In 1993, the French Ministry of Industry conducted an exhaustive survey (the survey "Globalisation") of all the firms located in France in order to determine the portion of foreign trade corresponding to international exchanges within multinational enterprises.

The purpose in this chapter is to exploit the results of this survey in order to explore the determinants of intrafirm trade carried out by multinational agribusiness. From an empirical point of view, this question is all the more important, as France is one of the world leaders in agrofood product exports. In 1998, it was the second largest food-product exporter after the United States and was the first exporter in Europe. Three-quarters of these French foreign exchanges are carried out by multinational firms (MNFs) that have subsidiaries abroad, and 23 percent of these exports are carried out in their internal market.

From a theoretical point of view, these intrafirm exchanges raise the question of the internalization of the market within the firm. It raises the question of the existence of a market internal to the multinational firm that would function with a different logic from that of the external market. In fact, most analyses are based on the assumption that the determinants of intrafirm exchanges are different from those of arm's-length transactions. However, with the exception of Benvignati's study (1990), no work has really tested this difference. The results obtained by this author do not enable one to conclude that there is a significant difference in the case of American multinational firms. The aim of this chapter is to reach an understanding of why enterprises export via their internal market and whether these determinants are different from those of traditional exports.

The next section of this chapter presents the theoretical framework and the empirical literature review in which the determinants of intrafirm trade are discussed. Subsequent sections present the empirical framework of the analysis of French multinational agribusiness. First, we consider the importance of intrafirm trade in the French food industry; second, we present the model and empirical results of the estimations.

THE FOUNDATIONS OF INTRAFIRM TRADE

Because a multinational firm is present in several countries and controls foreign affiliates, it is an organization that internalizes part of its activities at the international level. This internalization is the starting point for adapting the theory of the firm to the multinational enterprise by the neoinstitutionalist approach of transaction costs. This interpretative approach, centered on a positioning of the organization in relation to the market—i.e., organization as a response to "market failures"—can be enriched by a more internal approach of organizational mechanisms as a basis of internalization of transactions at an international level.

Two main approaches in economic literature have shown particular interest in this internalization theme. The first one, based on the theory of transaction costs and supported by O. E. Williamson (1975), distinguishes two institutions: the market and the firm. It stresses the fact that internalization responds to the need to avoid the costs that are inherent to exchanges on the market. The market failures, the barriers to international trade, and the resulting costs explain the internalization of the market within the multinational firm.[1] The second paradigm, developed by Dunning (1981, 1988), proposes a more "eclectic" approach to multinationalization, organized around three necessary conditions for the firm's internationalization: ownership-specific advantages, location, and internalization. In this respect, the internal market that emerges following direct investments abroad enables the firm to benefit from both its specific advantages and those related to location. Thus with respect to the foundations of internalization, the literature highlights three main types of determinants that refer to three basic functions of the firm: industrial, commercial, and financial.

1. The internalization of exchanges enables a multinational firm to not only reduce the uncertainty and costs related to transactions on

the market but also to protect and valorize its specific advantages, such as those related to products (technological property, trademarks, know-how) or to its organizational competencies (economies of scale and competence, experience transfer). These specific advantages, managed internally, generate intrafirm exchanges and provide multinational firms with a greater response capacity, not only because of special access to cross-border production capacity but rather because of information advantages arising out of multinational operations (Rangan and Lawrence, 1999). Thus internalization translates the existence of a networked international organization that is likely to facilitate the growth of multinational firms not only because markets are larger, as usually argued, but also because firms will be able to specialize their affiliates and launch new forms of international organization (Rama, 1998; Cantwell and Randaccio, 1993). Thus the role of specific industrial and organizational characteristics of the firm is considered in the literature as promoting intrafirm exchange. In this context, several explanatory factors are used:

- The search for *economies of scale* is believed to be a factor influencing the development of intrafirm trade. Markusen (1995) distinguishes the economies of scale obtained at the level of the corporate group from the economies of scale generated at the affiliate level. According to Markusen, the first type of economy specifically generates intrafirm transactions, so as to valorize research and development activities, for example. He refers to the economies of scope highlighted by Chandler (1990) that can be accounted for through *the international experience acquired* by the multinational firm. Thus, as shown by Siddarthan and Kumar (1990), if a corporate group already has affiliates in a country, its exchanges with this country will be intrafirm trade, so as to minimize the costs related to market information and to the creation of a new infrastructure to distribute the products. Authors such as Wang and Connor (1996) evaluate this experience by the percentage of sales made abroad by the affiliates, out of the total sales of the corporate group. In a similar vein, Lall (1978) makes use of the share of the assets of the firm abroad out of its total assets. The second type of economies of scale, which is internal to the affiliate, refers to the existence of increasing returns and of productivity gains related to the size of the firm. It explains the location of a plant in the given country by the necessity to achieve returns on equipment investments and to manufacture a product that corresponds to the demand of the host country. In order to evaluate these economies, Wang and Connor (1996), as well as Benvignati (1990), estimate an index of "minimal efficient size" of enterprises on a given market, based on the distribution of the turnovers or on the number of employees per sector of activity.

• According to Wang and Connor (1996), Siddarthan and Kumar (1990), and Benvignati (1990), *the differentiation and the creation of new products* are central to the explanation of the internalization of exchanges. Indeed, it is believed that these internally exchanged products are new and complex products for consumers and the costs related to their marketing are high (marketing promotion, need for consumer information or an after-sale service). These characteristics imply the existence of substantial R&D activity. Because of the greater complexity of the products and the necessity to protect technological knowledge, the intensity of R&D is an argument in favor of intrafirm trade (Vernon, 1966; Markusen, 1995; Teece, 1985). As Michalet noted (1985), the protection of technology and the know-how that accompanies it will be ensured by the creation of an internal market within the firm.[2] In this context, the *intensity of the human capital* is often used to evaluate the extent of the knowledge that must be protected, and research and development expenditures are used to determine the nature of the innovation intended for the products. *Advertising expenditures* can also give an indication of the effect of product differentiation on intrafirm trade. According to Wang and Connor (1996), it is a tool that efficiently measures the "differentiation of enterprises" which use, in particular, their image to promote their products for export. However, the part played by advertising expenditures is interpreted differently by Benvignati (1990). For this author, advertising—which makes it possible to reach consumers directly— cancels the advantages of intrafirm trade. Lall (1978) also justifies the negative correlation between advertising expenditures and the intrafirm export rate by the fact that the former are necessary to sell mass-produced goods directly to consumers.

• Finally, the purposes of intrafirm trade (IFT) are thought to differ according to *the nature of the product,* that is to say, according to whether the goods are assigned for final consumption or for production. The fact that the exchange concerns an intermediate good rather than a processed product is an argument in favor of intrafirm trade that Siddarthan and Kumar (1990), Wang and Connor (1996), and Fontagné, Freudenberg, and Ünal-Kesenci (1999) have all highlighted. Intrafirm exchanges of intermediate goods are believed to indicate a vertical production organization on an international scale. However, Andersson and Fredriksson (2000) show that it is inaccurate to view IFT as provision of intermediate goods and that trade of both product categories is likely to be internalized for different reasons. The basic hypothesis is that trade in intermediates follows from vertical integration (the affiliate manufactures intermediate goods that it sells to another affiliate of the group for later conversion while trade in final products relates to horizontal integration (most of the affiliate's production is sold directly in the foreign country). The trade of processed goods that prevail in agribusiness constitutes an important component of intrafirm trade. Cantwell (1994) shows, for instance, that parallel to vertical integration that generates intrafirm trade, horizontal integration leads each affiliate to produce certain varieties of products for the local market and for export, and to import other varieties produced by other affiliates. These internal exchanges of finished goods are all the more important, as they necessitate a phase of adaptation to the local market and an after-sale service.

2. Internalization is also considered to be a way to create or keep the *monopolistic advantage,* or even to raise barriers to a market entry (Hymer, 1960; Mathieu, 1997). Internalization offers the means to practice different competitive strategies, or "anticompetitive" strategies as Dunning called them (1981), against competitors (predatory prices, price discrimination). In a similar vein, the existence of tariff or nontariff trade barriers justifies the existence of intrafirm trade in a situation of uncertainty (Jacquemin, 1989; Markusen and Maskus, 2001a; Becuwe, Mathieu, and Sevestre, 1998). In this way, as Greenaway (1993) noted, foreign direct investments are used as "bridgeheads" to bypass certain trade barriers and thus promote intrafirm trade. As a result, many authors have paid particular attention to the analysis of market structures (degree of concentration, size of firms) and have introduced the level of concentration of the sector-based

activity to evaluate this factor (Connor, 1983; Sugden, 1983; Wang and Connor, 1996; Galliano, 1995).

3. Finally, the foundations of intrafirm trade can also be found in *the overall financial management of the corporate group*. Most authors highlight the possibility for corporate groups to maximize their profit owing to the internalization of transactions (see for instance, Wang and Connor, 1996). A multinational firm can arbitrate between its objectives of market shares and those related to its margins. It can distribute activities and profits among affiliates according to the economic conditions that prevail in the host country (Harris, 1994; Cantwell, 1995) while taking into account management-related problems and incentives offered to managers (Donnenfeld and Prusa, 1995). The argument is also that IFT may ease FDI when there is a threat of possible confiscatory taxation by the host government, particularly under incomplete information (Konrad and Lommerud, 2001). This conception grants the firm the ability to fix internal prices (transfer prices) according to its market strategy or in order to obtain an overall minimization of taxes in a context where taxation systems are heterogeneous. Therefore, international intrafirm trade is a necessary condition for a transfer of revenue (Harris, 1993; Jacob, 1996). One can link this type of determinant to currency risk management. Through internalization, multinational firms are able to minimize the costs of transactions by choosing the most advantageous currency for invoicing and payment. However, although most authors highlight the importance of transfer prices in accounting for the components of financial management, few empirical studies introduce this variable because of a lack of available data.

The nature of exchange, as in the case of taxation problems, raises the question of the characteristics of the destination or origin countries of the intrafirm exchange. Specific advantages are related to the choice of location made by taking into account trade barriers, the tariff or taxation policies of the host country, factor costs, and the size and evolution of the markets. The firm can reduce the uncertainty concerning the external environment by organizing around its internal market, its own information barrier against both the competitors and the governments of the host countries. Internalization enables one to bypass the national public policies (quotas, customs barriers, price control, and taxes). In this respect, the choice of location of foreign direct investments and the characteristics of the host countries play a significant part in the intensity of intrafirm trade.

INTRAFIRM TRADE IN FRENCH AGROFOOD FOREIGN TRADE

The statistics of international trade do not distinguish intrafirm trade from exchanges between independent entities. For this reason, in 1994, the statistical services of the French Ministries of Industry (SESSI) and of Agriculture (SCEES) and the National Institutes of Statistics (INSEE) conducted an exhaustive survey called "Globalisation" which provided detailed informations about international intrafirm trade of all the enterprises located in France (flows in value and quantity per enterprise, per product, and per destination or origin).[3] It is this survey that we have used for the following sections of this chapter. Several other statistic files have been used to allow a comprehensive analysis of this survey. In particular, we have used:

- The file concerning the exporting and importing enterprises of the directorate general of French customs. This file enabled us to compare intrafirm trade and arm's-length transactions. It provides information on the flows of exchanges of enterprises (internal and external), while taking into account their markets, their products, the quantities exchanged, and the prices charged. This latter variable is important because, from a theoretical point of view, its meaning is thought to be different concerning the internal market of the firm (transfer prices, transaction prices), although few empirical studies have tested it due to a lack of data.
- The file of the annual surveys of enterprises (EAE, INSEE) that provides results on turnovers, workforce, wages, export rate, and advertising expenditures.
- The survey concerning the financial links between firms (LIFI, INSEE) that enabled us to reconstruct the boundaries of the coporate group and to identify the parent company.

On this basis, several observations can be made concerning intrafirm trade in the French agrofood industry.

The Dominant Role of French Business Groups in Foreign Trade

The surplus of the French agrofood trade balance, often attributed to external competitiveness, depends in fact more particularly on the firms organized into groups of companies. These corporate groups represent 86 percent of the sector's exports (see Table 6.1).

Their participation is considerably lower in terms of imports (41 percent). This performance can be explained by the part played by corporate groups with French capital that export more than foreign MNEs or independent firms. In this context, each of these three actors play a specific role in the French balance of trade (see Table 6.1). French corporate groups occupy a major position inasmuch as they represent 63 percent of exports and only 22 percent of imports, and they achieve a high trade surplus. On the other hand, foreign firms located in France are more directed toward import than French groups and over half of all their exchanges are controlled by noncommunity groups (American and Swiss), the rest being controlled by Dutch, Italian, and German groups. As for independent firms, they play a very different role in foreign trade. They carry out almost two-thirds of all agrofood imports (59 percent) but only 14 percent of total ex-

TABLE 6.1. The protagonists and the modes of French foreign exchanges of agrofood products.

Exports ($20.47 billion)		Imports ($16.35 billion)	
Sales	**Percentage**	**Purchases**	**Percentage**
To third parties	77	From third parties	89
Share of independent firms	14	Share of independent firms	59
Groups, including	63	Groups, including	30
French groups	49	French groups	19
Foreign groups	14	Foreign groups	11
To affiliates*	23	From affiliates*	11
French groups	14	French groups	3
Foreign groups	9	Foreign groups	8

Sources: Globalisation survey, 1993 (SESSI, SCEES, INSEE) and French Customs file.
*Intrafirm exchanges.

ports (see Table 6.2); thus their external balance is negative ($–6.8 billion).

The Importance and the Nature of the Internal Market

The organization advantage, on which groups depend, results from a strategy of market conquest based on direct investments abroad. Particularly important in this sector in the past few years, these investments allow the development of local production and generate transactions of various natures within the multinational firm. However, the recourse to the internal market is not systematic in the agrofood industry. The globalization and customs files enabled us to identify 317 multinational groups (566 affiliates) involved in the foreign trade of the sector. Out of these 317 groups, only half carry out intrafirm exchanges, but these 168 multinational firms have the biggest impact on external exchanges. They carry out 78 percent of the total exports and 36 percent of imports of French agrofood products. Their internal markets channel almost 30 percent of the total exchanges of the groups (exports plus imports) which represents, in fine, 23 percent of French exports and 11 percent of imports. Finally, these intrafirm exchanges have a positive impact at a national level inasmuch as they generate a trade surplus of $2.93 billion, i.e., more than two-thirds of the surpluses of the French balance of trade (see Table 6.2).

TABLE 6.2. Groups in the French agrofood external balance, in 1993.

Groups	Balance in billions of dollars
French groups	9.1
Intrafirm exchanges	2.35
Foreign groups	1.82
Intrafirm exchanges	0.58
Independent firms	–6.8
Total	4.12

Sources: Globalisation survery, 1993 (SESSI, SCEES, INSEE) and French Customs file.

The importance of the internal market of multinational firms varies according to the products, their origin, and destination. Three-quarters of intrafirm exchanges in the food-processing industry are exports, carried out mainly by French (61 percent), American (17 percent), and Dutch (10 percent) groups (see Table 6.3). These exports concern beverages, products included in the "other food products" category (sugar, chocolate, coffee, condiments, crackers, and biscuits), or dairy products. On the other hand, intrafirm imports are carried out by foreign groups (70 percent) (American, Dutch, and Swiss) and concern the same products as for export. Depending on the branch, the share of exchanges flowing within multinational firms can be important; it represents 35 to 45 percent of external exchanges in the case of more standardized and less perishable products, such as beverages, flour, processed cereal, canned vegetables, and "other food" products.

Because of the nature of the products and of the processes of production, the internalization of exchanges in the food-processing industry is less advanced than in the manufacturing industry. In the industrial sector, intrafirm trade represents 34 percent of exports and 31 percent of imports (Hannoun and Guerrier, 1998). The difference is most marked at the level of import behavior of foreign groups: 31 percent of industrial imports are carried out within the foreign groups against only 8 percent in the food-processing industry. So, does the comparative advantage of France in terms of food production hinder the intrafirm trade of foreign groups? Or is it rather the constraints of adapting to the consumer preference for variety that limit their weight? This could explain why the local production of foodstuffs by affiliates of foreign groups has become so important. In 1993 it was superior to the total value of the national imports of the branch. This shows that the debate on the impact of the many aspects of globalization, at the level of national economies, is not limited to trade.

The Single European Market: A Privileged Destination of the Intrafirm Trade of Agrofood Products

The analysis of the geographic distribution of exchanges according to the nationality of the groups is rich in information concerning the purposes of intrafirm trade and reveals certain mechanisms of integration of the agrofood markets in Europe. This process is already

TABLE 6.3. Intrafirm trade in the agrofood sector, according to the nationality of the importing group and the geographic destination.

Nationality of the importing groups	Distribution of intrafirm imports according to the geographic origin (in %)			"Intensity" of intrafirm import of the groups according to the geographic origin (in %)*		
	Total	EU	Rest of the world	Total	EU	Rest of the world
All groups	100	81.48	18.52	23.71	25.18	18.84
French	30.00	18.96	11.04	12.87	10.71	19.7
Dutch	16.87	15.26	1.61	42.19	44.16	29.67
German	8.16	7.49	0.67	48.58	48.48	49.79
Italian	4.93	4.41	0.52	22.97	37.53	5.33
Other EU	5.96	5.83	0.13	46.61	49.82	12.76
Swiss	15.07	12.98	2.09	47.55	51.04	33.36
American	19.01	16.55	2.46	28.79	34.73	13.38
Rest of the world	0	0	0	0.94	1.81	0.71

Sources: Globalisation survey, 1993 (SESSI, SCEES, INSEE). French Customs File.
*The intensity of intrafirm imports of a group is the share of the intrafirm imports in its total imports (or for a given geographic origin). Thus, intrafirm imports represent 12.87 percent of total imports of French groups; 19.7 percent of the imports originating in the rest of the world.

well on the way to completion; it rests mainly on one of the basic principles of the common agricultural policy, that of community preference. The polarization of the groups' exchanges on the European Union (76.7 percent of imports, 67 percent of exports) testifies to this. Moreover, despite the lowering of trade barriers, most intrafirm exchanges are carried out within the single market: 81.5 percent of intrafirm imports originate in an affiliate located in Europe and 71.7 percent in the case of exports (see Table 6.4).

If the completion of the single market has led to the total liberalization of the capital market at the intracommunity level, as well as to the lowering of trade barriers, it also implies a common protection at the borders of the union. This mechanism is a traditional incentive for direct investments from third-party nations. These bridgehead investments (as Greenaway called them in 1993) enable one to bypass the barriers at the entrance of the single market and to have access to a demand that is even larger than the French demand alone. Thus most exchanges of the French affiliates of non-European groups (American and Swiss mainly) occur on the European market (72 percent for imports, 83 percent for their exports). Conversely, the small volume of trade of these affiliates with the rest of the world (and with the United States in particular) demonstrates that they are productive and not just commercial affiliates. The products imported from the mother company are not resold on the European market. This European penetration is partly based on the internal market of foreign groups and therefore on a network of affiliates that are also located in Europe (39 percent of exports to the European Union by French affiliates of American groups are intrafirm, 53 percent in the case of Swiss affiliates).

Furthermore, far from reducing exchanges between European firms, the single market promotes the construction of their own regional networks. Thus it is for their imports that the French affiliates of European groups use the internal market most (with intrafirm import rates reaching almost 50 percent in the case of German groups). Linked to the fact that most products are resold unmodified, intrafirm trade enables firms to offer more comprehensive ranges of products, thereby satisfying consumers' preference for variety. The internal market is used as a means to exploit the firm's specific advantages in a single market characterized by a process of convergence of economic structures. In this context, European groups, just as noncommu-

TABLE 6.4. Intrafirm trade in the agrofood sector, according to the nationality of the exporting group and the geographic destination.

Nationality of the exporting groups	Distribution of intrafirm export according to the geographic destination (in %)			"Intensity" of the intrafirm exports of the groups according to the geographic destination (in %)		
	Total	EU	Rest of the world	Total	EU	Rest of the world
All groups	100	71.75	28.25	25.38	27.12	21.81
French	60.72	41.96	18.76	21.19	22.57	18.66
Dutch	9.39	4.27	5.12	46.35	39.57	54.07
German	0.66	0.54	0.12	13.21	13.04	14.12
Italian	4.68	4.2	0.48	30.04	34.35	14.34
Other EU	0.83	0.69	0.14	13.33	18.35	5.71
Swiss	7.19	5.86	1.33	44.44	53.19	25.76
American	16.5	14.22	2.28	37.62	38.95	31.00
Rest of the world	0.03	0.01	0.02	6.15	5.18	6.48

Sources: Globalisation survey, 1993 (SESSI, SCEES, INSEE) and French Customs file.

nity foreign groups, exploit the advantages of the single market while contributing to the very process of integration in return.

THE DETERMINANTS OF INTRAFIRM TRADE IN THE FRENCH AGROFOOD INDUSTRY: MODEL AND RESULTS

The purpose of the analysis presented in this section is to explore the determinants of intrafirm trade carried out by multinational agribusiness and to verify whether the determinants of the volume of direct exchanges have a different impact from those that influence the volume of internal transactions.[4] Thus the model to estimate is of the following type:

$$IX = \beta_0 + \gamma_1 Z_x Intra + \gamma_0 Z_x (1 - Intra) + \delta Intra + u \qquad (6.1)$$

where *IX* is the volume of exports, Z_x is a vector of explanatory variables of *IX*, and *Intra* is a dichotomous variable which is equal to 1 if the exchange is an intrafirm exchange, and 0 otherwise. The level of observation is the triplet (enterprise, product, market) where

- the enterprises are the food-processing affiliates of groups that participated in the survey;
- the product is apprehended at the most detailed level of the customs nomenclature (eight-digit combined nomenclature, that is here 884 products concerned); and
- the geographic zones or markets retained are the eleven member states of the EU in 1993, Switzerland, the United States, Japan, and the rest of the world.

However, before carrying out a more thorough examination of this hypothesis, it is necessary to understand the choice made by a firm to not always use the internal market. The motivations of this decision must be taken into account to appreciate the impact of the use of the internal market on the volume of export. This has justified the choice of the estimation method following a Heckman procedure in two stages to obtain nonbiased estimators (see Box 6.1). However, prices which are an explanatory variable of the export function, are also one of the elements of the export strategy of a firm, whether or not the

BOX 6.1. The Econometric Model of the Export Function

Our econometric model is intended to test whether determinants are different according to whether the exchange is carried out on the internal or the external market. Thus the model we estimated is as follows:

$$lx = \beta_0 + \gamma_1 Z_x Intra + \gamma_0 Z_x (1 - Intra) + \delta Intra + u \qquad (6.2)$$

where lx is the export volume, Z_x is a vector of indendent variables, and *Intra* is a dummy variable (*Intra* = 1 if the exchange is an IFT; *Intra* = 0 otherwise).
Before estimating this equation, two problems must be solved.

Taking into Account the Underlying Selection Process: The Heckman Procedure

The export volume is a positive or null, truncated variable. The zero values may be explained by the decision of the firm to export within or outside the MNF. Consequently, there is a selection bias. In order to correct this bias, we have used an Heckman procedure (corrected by Greene, 1997). This is a two-step estimation:
In the first step, we estimate the following probit equation

$$Intra = 1 \text{ si } \hat{\alpha} Z + 3 > 0 \qquad (6.3)$$
$$Intra = 0 \text{ sinon}$$

This step leads to the estimate of $\hat{\alpha}$ and for each observation of the selected sample, it is possible to compute the Mills ratio: $M = \dfrac{\varphi(\hat{\alpha} Z)}{\Phi(\hat{\alpha} Z)}$ if *Intra* = 1 where $\hat{\alpha}$ is the prediction of the probit, φ the probability density of the normal law, and Φ the function of cumulated probability.
In the second step, i.e., the export volume regression, the Mills ratio is introduced along with the other explicative variables.
Equation 6.2 then becomes

$$lx = \beta_0 + \gamma_1 Z_x Intra + \gamma_0 Z_x (1 - Intra) + \lambda_1 M \, Intra + \lambda_0 M (1 - Intra)$$
$$+ \delta Intra + u \qquad (6.4)$$

Taking into Account the Endogeneity of the Price Variable: Method of Instrumental Variable and Hausman Test

Among the variables Z_x, those related to the prices may be endogenous to the model. In fact, explicative variables of the prices may also explain the export volume, in particular the differential of tax rates. Thus the explicative variable may be correlated with the error term *u*. Therefore, estimates are biased.

> In order to correct this bias, we have used an instrumental variable method, by using a 2SLS procedure and by estimating the following model:
>
> $$\begin{cases} Ix = \beta_0 + \gamma_1 X_z Intra + \gamma_0 Z_x (1 - Intra) + \lambda_1 M \, Intra + \lambda_0 M (1 - Intra) \\ + \delta Intra + u \\ Ip_x = \phi_0 + \vartheta_1 X_{px} Intra + \vartheta_0 Z_{px} (1 - Intra) + \phi Intra + e \end{cases} \quad (6.5)$$
>
> where Z_{px} is a vector of explanatory variables of I_{px} (the price).
> Finally, we have performed a Hausman test to determine whether a simple OLS estimation would have been consistent for our model.

exchanges occur on its internal market. In this respect, it was necessary to test the possible endogeneity of prices in the model, by using the method of instrumental variables and by performing a Hausman test to evaluate this endogeneity (see Box 6.1).

This procedure enables one to understand both stages of the export process (choice of the internal or external market and factors influencing the volume of exports) as well as the process of formation of intrafirm supply prices.

In keeping with our hypotheses, the explanatory variables are relative to the nature of the products and of their markets on the one hand, and to the characteristics of the firm's affiliates and the group it belongs to on the other. These variables are defined in Table 6.5.

The Determinants of the Choice to Use the Internal Market

Multinational firms can choose between an arm's-length transaction or selling to affiliates in foreign markets. How do they make this choice? In light of the previous literature review, we expect an MNF's decision to send goods to an affiliate abroad to be positively influenced by factors related to

- economies of scale,
- the concentration of the market in the activity considered,
- the international experience acquired by the group in foreign markets,
- the part of the budget dedicated to advertising,
- the average level of qualification of the firm's workforce,

- the presence of the group's mother company in the destination country,
- the tax rate differential between France and the destination country, or even
- the intermediate or finished nature of the goods.

The modeling of the group's decision (see Table 6.6) to choose an internal exchange reveals that most coefficients of the estimations are significant and that their signs seem to be in keeping with theoretical expectations. As Wang and Connor (1996) highlighted, the economies of scale at the firm level have a positive impact on the probability to carry out intrafirm exchanges. This factor is reinforced by the corporate group's international experience which reflects its know-how and the intensity of its involvement on international markets. This result is in keeping with those of Markusen (1995) who noted the significant influence of the group's experience on intrafirm transactions. Although authors are divided regarding the effects of advertising expenditures on intrafirm trade, we find here that they have a positive impact on the decision to use the internal market. Therefore, at this level, the argument could consist of the consideration of the role of advertising expenditure as a branding investment and as an important element of the construction of the firm's identity. This effect, which is favorable to the decision to use the internal market, is accentuated by the level of qualification of the firm's workforce, which (although it is only approximately covered by our indicator) seems to point to the existence of a know-how that must be protected. Finally, the concentration of the markets, which is believed to act as an exchange barrier, promotes intrafirm exchanges.

Moreover, in the case of multinational firms, which by definition have a productive organization at the international scale, we would expect intermediate goods to be more particularly destined to an affiliate of the group. On the contrary, it turns out that the intermediate nature of the goods has a negative effect on the decision to sell to an affiliate. In fact, what we are dealing with here is a characteristic of the food-processing sector in which there are two kinds of intermediate products: standard intermediate goods (lactoserum, for example) and strategic intermediate components used in the realization of specific finished products (Coca-Cola's flavorants, for example). In the latter case, one could assume that using the internal market enables a firm

TABLE 6.5. Definitions of the variables.

Variables	Measure
X_{ijk}	Quantity of export of the firm (j), an affiliate of the group, for the product (i) sent to the zone (k). The products are observed in nc8 nomenclature.

Characteristics of the exchange and of the market

Price	Unit value (value/quantity) for the product (i) exported by the firm (j), affiliate of the group
Degree of concentration of the export market	Herfindahl index of concentration calculated on the basis of the total exports of the branch (defined in cpf6 French nomenclature)
Zone of exchange	Zone (k) of the products' destination: the eleven member states of the EU in 1993, the United States, Switzerland, Japan, and the rest of the world.
Taxes	Ratio of the taxation rate on corporations between the destination country and France
Intermediate goods	BI = 1 the product is an intermediate good, BI = 0 it is a final good
INTRA	Dichotomous variable characterizing the exchange, INTRA = 1 if the exchange is intrafirm, INTRA = 0 otherwise

Affiliate-specific variables

Economies of scale	Ratio of the firm's size to the average size of the top 50 percent of firm size distribution of the same industry. The industry is defined from the French nomenclature (naf 600).
Wage ratio	Ratio of the average wage rate of the firm to the sector's average wage rate
Advertising expenditures of the affiliate	Part of advertising expenditures in the turnover (before tax) of the firm
Productivity	Ratio of the value added of the firm to the total number of employees
Firm's openness rate	Part of exports in the turnover (excluding tax) of the firm

Group-specific variables

Advertising expenditures of the group	Part of advertising expenditures in the turnover (excluding tax) of the group to which the firm belongs
Openness rate of the group	Part of exports in the turnover (excluding tax) of the group to which the firm belongs (j). This ratio is calculated without taking the affiliate into account.
Location of the parent	Dichotomous variable taking into account the nationality of the parent controlling the firm (j), Loc = 1 if the export is destined to the country of the parent, Loc = 0 otherwise.

TABLE 6.6. Factors influencing intrafirm trade (analysis of the probit).

Intra	Coefficient	P > |z|
Characteristics of the firm		
Economies of scale of the firm	0.216	0.000
Wages ratio	0.2296	0.047
Characteristics of the group		
Openness rate of the group	0.280	0.000
Advertising expenditures of the group	0.157	0.000
Location of the group's mother company	0.713	0.000
Nature of the goods or of the market		
Intermediate goods	−0.158	0.002
Degree of concentration of the market	0.266	0.000
Destination of the exchange		
BLEU (Belgo-Luxembourg Economic Union)	1.583	0.000
Netherlands	1.023	0.000
Germany	1.626	0.000
Italy	1.204	0.000
Great Britain	1.425	0.000
Denmark	0.118	0.615
Greece	−0.335	0.206
Portugal	0.338	0.127
Spain	1.583	0.000
Switzerland	0.991	0.000
United States	1.209	0.000
Japan	0.651	0.004
Constant	0.582	0.000

Probit estimates: number of obs = 5,956; $chi^2(22)$ = 961.44; Prob > chi^2 = 0.0000; log likelihood = −2930.237; Pseudo R^2 = 0.1681. NB: The zones were tested in relation to Ireland, a country that is outside the EU's core and with which France carries out few exchanges of food products.

to protect the know-how of the group. But the data used do not enable us to distinguish between both types of intermediate products.

Finally, the destination of the exchange seems to have an impact on the decision to carry out intrafirm trade. In first analysis, the destination zones that are favorable to intrafirm trade appear to be the countries that were part of the core of the European construction (Belgium, the Netherlands, Germany, Italy, Great Britain, and Spain). Indeed, these countries constitute zones of convergence in terms of economy and of the structures of the demand, in particular of the demand for food (Chevassus-Lozza and Gallezot, 1993; Freudenberg and Lemoine, 1999; Markusen and Maskus, 2001b). For other European zones, the probability of carrying out intrafirm trade is not significantly different from that observed in the reference country of the model (Ireland). This result can be explained by analyzing intrafirm exchanges as a means of adapting to the more specific constraints of the local demand. Conversely, and in reference to the theories of endogenous growth, another explanation could be that intrafirm trade positively influences the convergence. The historical dimension of the intensive development of direct investment and of the affiliation of enterprises in this zone would then be an important factor of the constitution of this European core.

The Formation of Intrafirm Prices

The existence of an internal market presupposes that a multinational firm can fix selling prices that are different from those of the market. This consideration makes endogenous the role played by prices in the explanation of the volume of exports (see Box 6.1). Thus before examining the impact of intrafirm exchanges on the volume of export, it is necessary to explain how the firm determines its prices (Table 6.7).

It turns out that the determinants that traditionally affect prices are not different for the sales of multinational firms targeted at affiliates or directly exported. Thus the higher the productivity of the firm, the higher the influence on the reduction of the price. Furthermore, a high specialization—believed to be revealed by the degree of openness on foreign markets—has a positive impact on prices, without the influence of these factors being significantly different from the internal market and direct export.

However, taking taxation into account seems to have a discriminating effect. If taxation on corporations in France is comparatively higher than in destination countries, it will penalize the competitiveness of the firm by exercising an increasing influence on the price of the goods.[5] This constraint, while it has an impact on direct exports

TABLE 6.7. The determinants of export prices.

Prices	Nature of the exchange	Coeff.	P > |t|	Equality of the coeff.*
Productivity	Intrafirm	–0.185	0.004	NS
	Direct	–0.119	0.000	
Taxes	Intrafirm	0.050	0.135	***
	Direct	–0.038	0.025	
Rate of openness of the firm	Intrafirm	0.087	0.000	NS
	Direct	0.057	0.000	
Choice of the firm (Mills ratio)	Intrafirm	–0.131	0.017	
	Direct	–0.175	0.000	
Intra		0.170	0.028	
European Union		–0.222	0.000	
Constant		2.846	0.000	

Regression with robust standard errors
Number of obs = 5,973; R^2 = 0.6140; adj R^2 = 0.6043.
P > |t| indicates the probability of the student test for each coefficient; the equality of the coefficients is tested by the Fisher test; *NS means that the difference is not significant at the threshold of 10 percent; ***the difference is significant at the threshold of 5 percent.

for countries that are comparatively less taxed, does not come into play if the multinational firm sells to an affiliate. This suggests that, in the determination of the internal price, the multinational firm can cancel the effect of taxation which restricts its competitiveness on the destination market.

The Impact of Intrafirm Trade on the Volume of Exports

The analysis of the role of the internal market on the volume of exports of a firm takes into account both its decision whether to carry out intrafirm exchanges and the endogenous formation of the prices. The question is now to determine if the explanatory factors of exchange are different depending on whether the exchange occurs through the affiliates or directly in the market. The comparison of the respective influence of the variables explaining the exports enables one to highlight the elements of these differences (see Table 6.8).

The results obtained show that whether the exports are direct or intrafirm, the lower the prices, the bigger the volume of exports. However, this relation is significantly more intense on the internal market than on the external market. Having previously set aside the influence of a tax system on prices by using price instrumentation in the modeling, this higher sensitivity of the volume of internal exchange in relation to prices can suggest—without mentioning dumping—the existence of an overall strategy of the group trying to optimize its price competitiveness in order to gain market shares. This hypothesis is all the more justified as the exchanged products are mostly finished goods destined to be resold as they are. However, this

TABLE 6.8. The determinants of the volume of exports.

Volume of exports	Nature of the exchange	Coeff.	P > |t|	Equality of coeff*
Prices	Intrafirm	−1.164	0.000	***
	Direct	−0.827	0.000	
Openness rate of the group	Intrafirm	0.556	0.000	***
	Direct	0.344	0.000	
Advertising expenditures of the group	Intrafirm	0.122	0.000	***
	Direct	−0.028	0.131	
Degree of concentration	Intrafirm	−0.353	0.000	NS
	Direct	−0.159	0.003	
Mills ratio	Intrafirm	−0.351	0.049	***
	Direct	−0.201	0.281	
Intraconstant		1.605	0.006	
		5.520	0.000	
Instrumented variables	Intrafirm prices and direct prices			
Instruments	Taxes, productivity, and rate of openness of the firms			

(2SLS) regression with robust standard errors.
Numbers of obs = 5,808; $F_{(23, 5784)}$ = 65.70; Prob > F = 0.0000; R^2 = 0.2544.
P > \|t\| indicates the probability of the student test for each coefficient; the equality of the coefficient is tested with the Fisher test; *NS means that the difference is not significant at the threshold of 10 percent; *** means that the difference is significant at the threshold of 5 percent.

motive does not entirely cancel those related to possible transfers of resources, other than those related to taxation.

The other results highlight the part (enhanced by intrafirm trade) played by the internal organizational dimension and the specific advantages of multinational firms. The positive effect that the international experience of the group has on exports is significantly more important on internal markets than on direct exchanges. This variable has also strongly influenced the initial choice of using the internal market. Advertising is advantageous to intrafirm trade. We recall here the findings of Wang and Connor (1996) for whom advertising expenditures account for the consolidation of the image of the firm. Finally, the concentration on the markets does have the negative effect expected on the volume of exports, without being different on the internal market.

CONCLUSIONS

The importance of intrafirm trade reveals one of the many aspects of globalization and, in many respects, provides a new representation of the role of firms and trade in the context of the international economy. The possibility of multinational firms selling to affiliates of the group rather than to a third party is at the heart of this chapter. The aim of this chapter has been both to specify the determinants that influence the decision to use the internal market and to reach an understanding of the specific impact of these internal sales, in relation with the impact of direct sales, on the volume of French agrofood exports. We tested the hypothesis that the functioning of the internal and external markets depend on different variables. This approach necessitated detailed information concerning French foreign exchanges and their distribution among intrafirm trade and arm's-length transactions. This approach was possible owing to the use of an exhaustive survey carried out by the French Institutes of Statistics concerning the role of intrafirm trade in the foreign trade of the nation.

In terms of results, the econometric use of the French survey on intrafirm trade highlights the positive influence of the search of valorization of firm-specific advantages and of protection against competition. Thus the exploitation of economies of scale, the valorization of the firm's image and of its qualifications, or the international experience of the group have a positive impact on the decision to carry out intrafirm trade. The concentration of the markets also has a positive influence inasmuch as it increases the probability of choosing the internal market. The intermediate nature of the goods exchanged, which is specific to the agrofood sector, is the only element that has a negative influence on intrafirm trade.

Second, the results confirm that the functioning of the internal market of multinational firms is comparatively different from that of direct exchanges on the market. This result, which has the advantage of being based on individual data of firms, does not converge with Benvignati's previous research on the U.S. case. Thus we have shown that the determinants of transfer prices and their influence on the intensity of trade differ significantly from those of market prices. These price effects, coupled with the management by multinational firms of taxation differentials, suggest an effort of overall optimization in which the internal market plays a central part. Furthermore, the functioning of the market shows the organizational advantage of multinational firms inasmuch as it enables them to make better use of their economies of scale, to protect their trademark and know-how, and to bypass trade barriers. Thus the internal market provides a multinational firm with both a space to manage industrial and commercial strategies and an advantageous space for its financial management. Because of its ability to bypass the external market and national barriers, it raises the question of the impact and efficiency of public policies (exchange rates, trade policies) on national competitiveness.

NOTES

1. These barriers to international trade are many. In the context of a study concerning transaction costs, Casson (1985) carried out a specific analysis of sectors, such as the food industry, with low technological development and a greater interest in product quality and marketing skills. The obstacles included a lack of communication between buyers and sellers, ignorance of mutual needs, lack of agreement on price, lack of confidence that goods correspond to specifications, no restitution of goods in case of nonpayment, the need to transport the goods, and the existence of customs duties, taxes, price controls, and quotas. According to Casson, internalization constitutes one of the main means of ensuring product quality and reducing the costs related to negotiation and the establishment of trust when information is lacking.

2. In the food industry, this is all the more important strategically, as the process of innovation in this sector tends to be a cumulative process (past innovation

strongly influences current innovation) and then a key factor in the competitive strategy of global firms (Alfranca, Rama, and Tunzelmann, 2001; Rama, 1999).

3. The survey was conducted on all the industrial firms localized in France and belonging to a multinational group (French or foreign). The response rate exceeded 80 percent of the number of firms (6,800 enterprises, including 566 agrofood firms) and 95 percent of the total of MNF exports and imports in France.

4. On these topics, see also our previous works: Chevassus-Lozza, Gallezot, and Galliano, 1999a (which compared the determinants of imports plus exports but without taking into account the underlying selection process), 1999b.

5. The indicator concerning taxation is structured on the ratio between the taxation situation of the destination country and that of France. A reduction of this rate reflects a French taxation system that is less favorable than that of destination countries.

REFERENCES

Alfranca, O., R. Rama, and N. Von Tunzelmann (2001). Cumulative innovation in food and beverage multinationals. In D. Kantarelis (Ed.), Business and Economics Society International Conference 2001: Proceedings—*Global business and economic review, anthology 2001* (pp. 446-459). Worcester, MA: BESI.

Andersson, T. and T. Fredriksson (2000). Distinction between intermediate and finished products in intra-firm trade. *International Journal of Industrial Organization,* 18: 773-792.

Becuwe, B., C. Mathieu, and P. Sevestre (1998). Commerce intra-firme, coûts de production aléatoires et barrières aux échanges. *Revue Economique,* 49(3): 581-591.

Benvignati, A.M. (1990). Industry determinants and "differences" in U.S intrafirm and arms-length exports. *The Review of Economics and Statistics,* 72(3): 481-488.

Cantwell, J. (1994). The relationship between international trade and international production. In Greenaway, D. and Winters, A. (Eds.), *Survey in international trade* (pp. 303-328). Cambridge, UK: Basil Blackwell Press.

Cantwell, J.A. and S.S. Randaccio (1993). Multinationality and firrm growth. *Weltwirtschaftliches Archiv,* 129(2): 275-299.

Casson, M. (1985). Transaction cost and the theory of the multinational enterprise. In Buckley, P. and Casson, M. (Eds.), *The economic theory of the multinational enterprise* (pp. 20-38). London: Macmillan.

Chandler, A. (1990). *Scale and scope: The dynamic of industrial capitalism.* Cambridge, MA: Harvard University Press.

Chevassus-Lozza, E. and J. Gallezot (1993). L'intégration des marchés agroalimentaires européens. *Economie et Prospective Internationale,* 53: 81-94.

Chevassus-Lozza, E., J. Gallezot, and D. Galliano (1999a). Les déterminants des échanges internationaux intra-firme: Le cas de l'agro-alimentaire français. *Revue d'Économie Industrielle,* 87(1): 31-44.

Chevassus-Lozza, E., J. Gallezot, and D. Galliano (1999b). Exportations intra-firme ou directes: Une alternative pour les firmes multinationales. *Economie et Statistiques,* 326-327(6/7): 97-112.

Connor, J. (1983). Determinants of foreign direct investment by food and tobacco manufacturers. *American Journal of Agricultural Economics,* 65(2): 395-404.

Donnenfeld, S. and T.J. Prusa (1995). Monitoring and coordination in MNCs: Implications for transfer pricing and intra-firm trade. *Journal of Economic Integration,* 10(2): 230-255.

Dunning, J.H. (1981). *International production and the multinational enterprise.* London: George Allen and Unwin.

Dunning, J.H. (1988). The eclectic paradigm of international production: A restatement and some possible extensions. *Journal of International Business Studies,* 19: 1-31.

Fontagné, L., M. Freudenberg, D. Ünal-Kesenci (1999). Trade in technology and quality ladders: Where do EU countries stand? *International Journal of Development Planning Literature,* 14(4): 561-582.

Freudenberg, M. and F. Lemoine (1999). *Central and Eastern European countries in the international division of labour.* Communication, second Association for Comparative Economic Studies (EACES) Paris Workshop, March.

Galliano, D. (1995). *Les groupes industriels de l'agro-alimentaire français.* Paris: Inra/Economica.

Greenaway, D. (1993). Direct investment in EC. *European Economy,* 52: 113-141.

Greene, W.H. (1997). *Econometric analysis* (International edition, Third edition). Upper Saddle River, NJ: Prenctice Hall.

Hannoun, M. and G. Guerrier (1998). Les échanges intragroupe dans l'industrie. In *Industrie française et mondialisation* (pp. 121-128). Paris: SESSI.

Harris, D.G. (1993). The impact of U.S. tax law revision on multinational corporations' capital location and income shifting decisions. *Journal of Accounting Research,* 31(Supp.): 111-140.

Hymer, S.H. (1960). *The international corporations of national firms: A study of direct foreign investment.* Cambridge, MA: MIT Press.

Jacob, J. (1996). Taxes and transfer pricing: Income shifting and the volume of intrafirm transfers. *Journal of Accounting Research,* 34(2): 301-312.

Jacquemin, A. (1989). International and multinational strategic behaviour. *Kyklos,* 42(4): 495-513.

Konrad, K.A. and K.E. Lommerud (2001). Foreign direct investment, intra-firm trade, and ownership structure. *European Economic Review,* 45: 475-494.

Lall, S. (1978). The pattern of intra-firm exports by U.S multinationals. *Oxford Bulletin of Economics and Statistics,* 40: 209-222.

Markusen, J. (1995). The boundaries of multinational enterprises and the theory of international trade. *Journal of Economic Perspectives,* 9(2): 169-189.

Markusen, J. and K.E. Maskus (2001a). Multinational firms: Reconciling theory and evidence. In Blomstron, M. and Goldberg, L. (Eds.), *Topics in empirical in-*

ternational economics: A fest schrift in honor of Robert E. Lipsey (pp. 693-708). Chicago: University of Chicago Press.

Markusen, J. and K.E. Maskus (2001b). A unified approach to intra-industry trade and direct foreign investment. Working paper 8335. Cambridge, MA: NBER.

Mathieu, C. (1997). International enterprises and endogeneous market structure. *Annales d'Economie et de Statistique,* (July/September), 47.

Michalet, C.A. (1985). *Le capitalisme mondial,* Second edition. Economie en liberté series. Paris: Presses Universitaires de France.

Rama, R. (1998). Growth in food and drink multinationals, 1977-94: An empirical investigation. *Journal of International Food and Agribusiness Marketing,* 10(1) 31-52.

Rama, R. (1999). Innovation and profitability of global food firms: Testing for differences in the influence of home base. *Environment and Planning A,* 31: 735-751.

Rangan, S. and R.Z. Lawrence (1999). Search and deliberation in international exchange: Learning from multinational trade about lags, distance effects, and home bias. Working paper 7012. Cambridge, MA: NBER.

Siddarthan, N. and N. Kumar (1990). The determinants of inter-industry variations in the proportion of intra-firm trade: The behaviour of U.S multinationals. *Weltwirtschaftliches Archiv,* 126(3): 581-591.

Sugden, R. (1983). The degree of monopoly, international trade, and transnational corporations. *International Journal of Industrial Organization,* 1: 165, 187.

Teece, D.J. (1985). Multinational enterprise, internal governance, and industrial organization [papers and proceedings]. *American Economic Review,* 75(2): 233-238.

Vernon, R. (1966). International investment and international trade in the product cycle. *Quarterly Journal of Economics,* 91: 190-207.

Wang, K. and J. Connor (1996). The determinants of intra-firm international trade. In Sheldon, M. and Abbott, P. (Eds.), *Industrial organization and trade in the food industries* (pp. 162-181). Boulder, CO: Westview Press.

Williamson, O.E. (1975). *Market and hierarchies, analysis and anti-trust implications.* New York: The Free Press.

Chapter 7

The Internationalization Paths of Australian and New Zealand Food MNEs

Bill Pritchard

INTRODUCTION

Since the 1980s Australia and New Zealand have been at the global forefront of agrofood trade liberalization. Both nations have progressively deregulated their agricultural sectors, reduced border protection for agrofood imports, and lobbied intensely for multilateral trade reform. These policy shifts have been justified through recourse to economic models suggesting that domestic and multilateral trade liberalization would generate significant national economic gains. Australian and New Zealand governments have argued that creating an increasingly level playing field for agrofood industries would facilitate allocative efficiency in resource use and enhance competitive advantage. Implicit in these decisions has been an assumption that improved efficiencies at home and expanded export opportunities abroad would provide fertile terrain for antipodean[1] food multinational enterprises (MNEs). In other words, global agrofood liberalization would present opportunities for antipodean agrofood companies to develop *internationalization paths,* whereby the increased possibility for the export of products from Australia and New Zealand would be accompanied by investment strategies in third countries (Pritchard, 1999b).

I would like to acknowledge the research assistance of Rebecca Curtis in the preparation of this chapter.

This chapter critiques the suggestion that the domestic deregulation of Australian and New Zealand agrofood industries, in combination with general trends toward the liberalization of international agrofood trade, is encouraging the emergence of antipodean MNEs. Using an approach that applies Dunning's research on firm internationalization to three detailed case studies, it is argued that the internationalization paths of agrofood firms depend upon complex intersections of firm strategy and industry context. When analyzed in these terms, there are strong reasons to challenge assumptions of Australia and New Zealand as emerging home nations for agrofood MNEs. In turn, this raises questions about the capacity for internationalization from agrofood firms based in relatively small and open economies.

This chapter pursues these issues in the following way. First, existing literatures on firm internationalization paths are reviewed. The *industry-specific* influences on antipodean food firm internationalization are examined, emphasizing the historical and politicoeconomic factors that have shaped the evolution of these industries and created contexts for agrofood internationalization. In revealing both the limited extent of internationalization and the diversity of situations that have given rise to these processes, discussion then turns to an analysis of *firm-specific* issues. Dunning's "eclectic framework" is applied to the three most prominent "internationalizers" in the antipodean agrofood sector. As illustrated in Table 7.1, these three companies represent the three most significant examples of Australia and New Zealand acting as home nations for agrofood internationalization.[2] Notably, each of these case study firms has developed quite different internationalization paths. The first case study, Foster's Brewing, has attempted to build global brand presence through franchising arrangements. The second, Coca-Cola Amatil, has attempted to use Australia as a base from which to capture markets in the Asia-Pacific region. The third, Fonterra, has attempted to leverage its competitiveness in agrocommodity production (in the dairy sector) to a global scale.

These case studies and the related discussion of Australia and New Zealand's food industries illuminate the indeterminacy and complexity in the relationship between the roles of these nations as agroexporters and their roles as home nations for food MNEs. The prospect of global agrofood trade liberalization may suggest an expanded

TABLE 7.1. Firms with Australian and New Zealand food sector production greater than US$100 million, 2000-2001.

Company	Year ending	Revenue from Aust/ NZ food sector	Aust/NZ food sector revenue as a % of global revenue	Ownership	Key activities
1. Fonterra[a]	Nov. 2001	5,461	N/A	CP	Dairy
2. Goodman Fielder	June 2001	1,304	79	PC	Grains, baking, poultry
3. Foster's Group	June 2001	1,086	46	PC	Beer, wine
4. Nestle Australia	Dec. 2000	1,044	—	FP	Consumer foods
5. ConAgra	May 2001	1,015	—	FP	Meat, grains
6. Coca-Cola Amatil	Dec. 2000	979	52	PC	Soft drinks
7. Lion Nathan	Sept. 2000	870	98	PC/FP	Beer, wine, spirits
8. George Weston Foods	July 2001	835	100	PC/FP	Baking, milling
9. Australian Wheat Board	Dec. 2000	780	100	SA	Grains
10. Dairy Farmers Group	June 2001	714	100	CP	Dairy
11. Murray-Goulburn	June 2001	696	100	PU	Dairy
12. Mars	Jan. 2001	641	—	FPr	Confectionery, pet foods
13. Cadbury Schweppes	Dec. 2000	635	—	FP	Confectionery
14. Bonlac Foods	June 2001	623	100	PU	Dairy
15. Southcorp	June 2001	541[c]	91	PC	Wine
16. National Foods	June 2001	502	100	PC	Dairy, juices
17. Campbell/Arnotts	July 2000	468	—	FP	Consumer foods
18. Grain Pool WA	Dec. 2000	440	100	SA	Grains
19. Inghams	June 2001	408[c]	100	Pv	Poultry
20. AFFCO Holdings	Sept. 2000	379	100	PC	Meat
21. Ricegrowers Coop	Apr. 2001	372	100	CP	Rice
22. Simplot	Aug. 2000	371	—	FP	Processed vegetables

TABLE 7.1 (continued)

Company	Year ending	Revenue from Aust/ NZ food sector	Aust/NZ food sector revenue as a % of global revenue	Ownership	Key activities
23. Unilever[b]	Dec. 2000	359	—	FP	Consumer foods
24. Parmalat	Dec. 2000	359	—	FPr	Dairy
25. Alliance Group	Sept. 2000	351	100	PC	Meat, livestock
26. Barrter Enterprises	June 2001	331	100	Pv	Poultry
27. Costa's	June 2001	327	100	Pv	Fruits and vegetables
28. Kraft Foods	Dec. 2000	304	—	FP	Dairy, consumer foods
29. BRL Hardy	Dec. 2000	278	85	PC	Wine
30. CSR	Mar. 2001	270	100	PC	Sugar
31. Nippon Meat Packers	Mar. 2001	263	—	FPr	Meat
32. Teys Brothers	Mar. 2001	260	100	Pv	Meat
33. Finasucre	Mar. 2001	252	—	FP	Sugar
34. Heinz Australia	May 2001	251	—	FP	Consumer foods
35. Kellogg's	Dec. 2000	236	—	FP	Consumer foods
36. Orlando Wyndham	Dec. 2000	228	—	FP	Wine
37. Berri	June 2001	226	100	PU	Juices
38. Smith's Snackfood	Dec. 2000	216	—	FP	Consumer foods
39. Bindaree Beef	June 2001	213	100	Pv	Meat
40. Manildra Group	June 2001	204	79	Pv	Grains, milling, sugar
41. DB Group	Sept. 2000	203	100	FP	Beer, wines, spirits
42. Peters & Browne	June 2001	202	—	PU	Dairy, ice cream
43. Australian Dairy Corp.	June 2001	190	100	SA	Dairy

44. Riverina (Mitsubishi)	Dec. 2000	179	—	FP	Meats, grains
45. Sanitarium	June 2001	148	100	Pv	Consumer foods
46. Snack Foods	June 2001	139	100	PC	Consumer foods
47. Golden Circle	Dec. 2000	135	100	PU	Fruit processing
48. Bunge	Jan. 2000	132	—	FP	Grains, meat
49. Warrnambool Cheese	June 2001	132	100	PU	Dairy
50. McCain Foods	June 2001	131	—	FPr	Vegetable processing
51. SPC	Dec. 2000	130	100	PU	Vegetable processing
52. Sara Lee Tea & Coffee	June 2000	126	—	FP	Consumer foods
53. Guiness UD	June 2000	125	—	FP	Spirits, beer
54. Chiquita Brands	Dec. 2000	124	—	PC/FP	Fruit and vegetables
55. Mackay Sugar Coop	June 2001	108	100	CP	Sugar
56. Simeon Wines	June 2001	103	100	PC	Wine

Source: Author's own compilation based on several sources. (Complexities of how the food industry is defined and the difficulties of gaining financial data for private-owned and foreign-owned firms necessitate a series of assumptions and estimates being made. Data compiled using the IBISWorld listing of Australasian companies <www.ibisworld.com.au> was amended where the author judged data to be inadequate or misleading. Listing includes firms in the following IBISWorld categories: "manufacturing–food," some firms from "wholesale," and some from "agriculture, forestry, and fishing." Some definitional issues exist with respect to non–Australian/New Zealand owned companies [Nestle Australia and Nestle New Zealand are treated as separate entities in the IBISWorld list, as are Heinz Australia and Wattie's New Zealand]. Original IBISWorld data in Australian dollars has been converted by author to U.S. dollars using the average 2000-2001 exchange rate of 1USD = 0.51 AUD.)

[a]The revenue figure for Fonterra is global revenue, because of the recent incorporation of this entity (see discussion in this chapter), at the time of writing it has not published its first annual report.

[b]Author's estimate that food activities comprise 60 percent of Unilever Australia's revenue.

[c]Author's estimates based on industry knowledge.

PC = publicly listed company in Australia or New Zealand; PU = unlisted public company in Australia or New Zealand; Pv = private-owned company; CP = cooperative; FP = foreign-owned publicly listed company; FPr = foreign-owned private company; PC/FP = publicly listed company in Australia/New Zealand but with majority foreign ownership; SA = statutory authority.

role for Australia and New Zealand in the world's food system; however, should these developments materialize, they will not necessarily be accompanied by the emergence of antipodean food MNEs.

EXISTING RESEARCH ON FIRM INTERNATIONALIZATION

Business studies literature points to a range of motives encouraging the internationalization of firms, including market seeking, asset seeking, resource seeking, efficiency seeking, and escape motives. These categories can provide useful devices for explaining key processes. Inevitably, however, the individual circumstances of particular firms generate unique internationalization paths. Empirical studies into MNE internationalization have identified considerable diversity, which makes generalization difficult (Sklair, 1998). As Goodman (1997, p. 665) suggests, "International economic integration and global restructuring are conceptualized and empirically represented as the dynamic conjuncture of several world-scale processes of capitalist accumulation and competition . . . that operate *concurrently* yet differentially in the world economy" (italics in original).

These arguments suggest the utility of an approach that structurally accounts for diversity in MNE internationalization. To this end, the work of Dunning (1971, 1993, 1997; Stopford and Dunning, 1983) is particularly attractive. Originating in an attempt to seek middle ground within broader political debates on capitalism and MNEs during the 1960s, which saw political economists (e.g., Baran and Sweezy, 1966) contest the dominant models and assumptions of business analysts and organizational theorists (e.g., Aharoni, 1966), Dunning's key contribution was the insight that an understanding of MNE behavior requires an approach that is highly sensitive to the varying situations and competition attributes of firms. Dunning argued that structural market failures discriminate among firms and lead to imperfect market outcomes: "Such variables as the structure of markets, transaction costs and managerial strategy of firms then become important determinants of transnational economic activity" (Dunning, 1993, p. 76). The approach suggests that researchers should not seek a single, explanatory factor to account for MNE behavior but instead should understand that the reasons for MNE activity rest within a configuration of factors, set within unique landscapes of competition.

Thus, according to Dunning, firm internationalization is determined by a complex configuration of industry- and firm-specific influences. The internationalization paths of individual firms are products of these intersections, giving rise to great diversity.

These principles led Dunning to employ a threefold classification schema to explain how MNEs internationalize (Dunning, 1993). For internationalization to occur, three conditions need to be satisfied. The first set of conditions is what Dunning refers to as ownership-specific (O) advantages, such as innovations, intangible assets, human capital, governance factors, or the specific organizational flexibilities provided through multinationality. In other words, internationalization occurs only if firms have an asset or capability that can form the basis for offshore activities. Second, there need to be location-specific (L) advantages, meaning the reasons why economic activities should be shifted outside a firm's home country. These reasons may include reduced costs or market access: the justification for servicing markets by offshore rather than home bases. Third, there is a need for internalization-specific (I) advantages, referring to the incentives for firms to retain economic activities in-house. The example of the American breakfast cereal MNE Kellogg's illustrates how internationalization can be expressed in terms of these three conditions being met. Kellogg's internationalization can be understood as occurring (1) because the firm has O advantages in its possession of recipes and trademarks that give it a competitive edge in international marketplaces; (2) because of transport costs and historically important market access barriers, there are significant advantages in servicing non-U.S. markets from non-U.S. production (L); and (3) to ensure control and management over lucrative tangible and intangible property, there are I advantages in retaining ownership of assets.

Dunning's OLI framework has substantial appeal. It is dynamic, it accommodates a range of theoretical insights into firm behavior, and it provides a matrix to help explain the particularities of firm strategy and performance. The approach is not without certain limitations—as some critics have noted (see Dicken, 1998), the framework itself says little per se about the mechanics of profit generation nor does it provide an explanatory model for firm behavior—nevertheless it gives structure to the task of understanding the complex influences that underpin and direct MNE internationalization. This framework is relevant for this chapter because the economic landscapes of regulation and competition from foreign MNEs have circumscribed strategic arenas for antipodean food companies, meaning that international expansion strategies have often been tied to the possession of firm-specific factors, such as distinctive or unique competitive assets of intellectual property, privileged access to raw materials, or a strategic position within global commodity chains. However, prior to the consideration of these factors in the three case studies discussed in this chapter, it is necessary to discuss the overall industry-specific contexts that have faced these companies.

THE HISTORY AND POLITICAL ECONOMY
OF ANTIPODEAN FOOD SECTORS

Australia and New Zealand share a unique space within the world food economy. As relatively low-cost producers of an array of important agricultural products (including dairy, beef, wool, grains, and sugar), these nations have led the charge for multilateral agricultural trade liberalization. Yet historically their agricultural export competitiveness has not provided a launching pad for the emergence of food MNEs. In 2000, only two Australian/New Zealand companies were ranked in the world's 100 largest food and beverage (F&B) corporations: the Foster's Group (ranked 95th) and Goodman Fielder (ranked 98th) (Anonymous, 2001).

The relative paucity of globally ranked antipodean food sector MNEs begs inquiry and encourages consideration of the historical contexts of antipodean agrofood political economy. Although subject to many complex influences today, Britain's invasion and colonization of Australia and New Zealand still casts a long shadow over the character of these countries' agricultural and processed food industries. British political control saw Australia and New Zealand develop as agricultural suppliers to the motherland. This economic role continued well after the formal political separation of colonial ties. British imperial preference allowed Australia and New Zealand to retain agricultural export markets during the middle decades of the twentieth century, helping these nations avert the problems of agricultural protectionism and declining commodity prices that befell some competing nations, notably Argentina (Platt and di Tella, 1985). At the same time, the relatively late colonization of Australia and New Zealand meant they developed as urban societies with a British culture of processed food consumption. Through most of the twentieth century, antipodean diets (like their British counterparts) were heavily dependent on the standardized fare of canned and bottled foods within a relatively unsophisticated palate of meat and vegetables. As argued by the food historian Michael Symons (1982), the Anglo-Celtic food cultures of Australia and New Zealand discouraged artisanal or region-based cuisines from developing, at least until being challenged by the cuisines introduced by waves of (non-British) immigrants in the second half of the twentieth century. The dominant Anglo-Celtic food culture, combined with tariff-based industry protection policies, provided lucrative environments for British and American food MNEs to establish branch plant operations in Australia and New Zealand, typically in the form of multidomestic, standalone facilities (Rama, 1992).

Table 7.1, which lists Australian and New Zealand firms with local food sector activities exceeding US$100 million, illustrates the legacy of this confluence of British-oriented agrocommodity export demand and a protected local processed-food industry dominated by foreign, branch-plant MNEs. Table 7.1 differs from other published lists of Australian and New Zealand food companies, in that (1) it separates food from nonfood activities within each company and (2) it separates antipodean food sector activities from those in offshore locations. Consequently, figures for each company are the estimated revenues of local food sector production, whether for domestic or export markets.[3] Hence, Table 7.1 allows for analysis of the composition of the antipodean agrofood sector and the position of food MNEs within these industry contexts.

Twenty-three of the top fifty-six firms in Table 7.1 are majority owned by non-Australian or non-New Zealand parent firms. These foreign firms control approximately one-half of Australian food industry capacity (Instate, 2000). Foreign ownership is most obvious in branded product sectors such as consumer foods, confectionery, and processed fruits and vegetables. Aside from Goodman Fielder, a diverse food conglomerate with operations across a range of product categories, there are no significantly large wholly antipodean-owned firms in this area (the next largest antipodean-owned firm in this sector, Sanitarium, has revenues of only US$148 million and is ranked 45th). Foreign ownership of antipodean branded foods gathered pace in the 1990s, via a series of high-profile acquisitions of local firms, including Arnotts (by Campbell Soup, 1993), Edgell (by Simplot, 1995), Peter's Ice Cream (by Nestle, 1995) and Paul's Ltd. (by Parmalat, 1998). Because of the extensive presence of foreign-owned firms in this sector, there is high market concentration in many individual product segments. In Australia, approximately half of all industry turnover is owned by the twenty largest companies (Instate, 2000) and seventy-one of the 100 largest-selling brands in Australian supermarkets are foreign owned (Dasey, 2001).

The dominance of foreign firms in this sector has circumscribed the capacities for internationalization by local branded food firms and, in general, has had a deadening effect on industry innovation and exports. With few exceptions, foreign-owned MNEs in the Australian food sector have tended to adopt multidomestic, branch-plant structures with little interest in using Australia and New Zealand as a base for processed-foods exporting. At times, the Australian operations of foreign-owned MNEs have been prohibited from exporting based on head-office edicts. In the mid-1990s, for example, Kellogg's Australia was stripped of its Asian export markets by its head office in the United States, because of changes in global strategy (Pritchard, 2000). Despite the avowed commitments of some foreign MNEs during the 1990s to increase their Australian exports, growth has been modest. Over the years 1994-1998, Australian processed-food exports grew by just 1.8 percent, compared with 21 percent for the United States, 9.9 percent for Germany, and 9.2 percent for France. In these same years, Australian exports of *unprocessed* foods grew by 40 percent (Instate, 2000). Most foreign-owned branded food companies appear to continue to use Australia and New Zealand as "cash cow" markets for their mature products, with relatively minimal attention to investment. The management of intangible assets (trademarks and brand names) has provided a cornerstone for company strategy. Over time, there has been a steadily rising outflow of royalty payments for the use of trademarks registered with offshore parent companies (Pritchard, 1999b, 2001). These market conditions underline the difficulties facing attempts by domestically owned branded food companies to internationalize. Goodman Fielder, the company with the greatest potential internationalization capacity, remains highly focused on local markets in the wake of a series of relatively poor-performing offshore acquisitions. Other antipodean-owned firms in this area (companies such as Golden Circle, Berri, Snack

Foods, and SPC Ardmona) tend to have 100 percent of their food activities within Australia/New Zealand and face a highly competitive market dominated by larger rivals.

Conditions in agrocommodity export sectors differ considerably from those in branded product markets. For much of Australia and New Zealand's European history, sectors such as dairy, beef, and canned fruits were heavily dependent on British import demand. Companies in these sectors evolved in the knowledge that British preference schemes provided relatively secure export markets, at least until the United Kingdom's entry to the European community in the early 1970s. Under the lengthy era of post-1945 conservative governments in Australia and New Zealand, rural political interests were managed through extensive "orderly markets" regulation. Like other agroexport countries during this era (which Friedmann and Mc-Michael [1989] dub the "second food regime"), state-supported producer boards played central roles in the industrial organization of key export commodity sectors. The Australian Wheat Board held (and continues to hold) an export monopoly for Australian wheat (its monopoly on domestic wheat was deregulated in 1989). Until 2002, the Australian Dairy Board possessed statutory single-desk marketing rights over the key market of Japan. The Australian export sugar industry has been organized through the allocation of export licences and quotas. In New Zealand, dairy exports were controlled until very recently through the single-desk powers of the New Zealand Dairy Board. In the case of the most labyrinthine of these structures, in the Australian dairy industry, the effect of regulations was to blur price signals, cross subsidize domestic and export markets, and institutionalize inefficiencies among participants (Australia: Industry Commission, 1991). These institutional factors impacted critically on the emergence of agrocommodity-exporting MNEs.

Removal of statutory controls has created new industry structures for Australian and New Zealand agrocommodity exporting, with varying effects for firm internationalization. Fonterra, the largest antipodean food firm, is a direct product from the decision of the New Zealand government to allow a corporatized New Zealand Dairy Board to merge with local cooperatives (discussed in detail later in this chapter). In Australia, where dairy deregulation has taken a different path, industry structures are the outcome of a succession of mergers among regional cooperatives over fifteen years of policy reform. In the beef sector, market liberalization has provided strong incentives for economic actors to generate productivity improvements, effected mainly through economies of scale strategies. This has led to the demise of smaller meat-packing operations and their replacement by state-of-the-art facilities, which tend to be owned either by foreign investors (e.g., ConAgra, Mitsubishi) or by expanding, privately owned firms (e.g., Teys Brothers, Bindaree Beef).

The abilities of some antipodean firms to expand in agrocommodity sectors raises issues relating to the capacities to add value to Australian and New Zealand primary products. Although many antipodean agricultural commodities are internationally cost competitive, this does not necessarily translate into cost-competitive processed-foods production. In many instances, price competitiveness at the farm-gate is offset by higher costs in processing, transport, and packaging (Instate, 2000). With a few exceptions (notably dairy, beef, and wine), the Australian processed-foods sector does not possess the economies of scale advantages available to competitors in Europe and North America. This is starkly apparent in horticulture, where Australia's capacity to capture export markets for fresh and processed product is stymied by scale disadvantages (Australia: Horticultural Task Force, 1994). Transport costs, including those at the waterfront, are relatively high, owing to poorly integrated domestic transport infrastructure and the remoteness of Australia and New Zealand from key export markets (Australia: Productivity Commission, 1998). Packaging is relatively expensive compared to international benchmarks because, in the opinion of a number of food industry executives, the small size of the Australia-New Zealand market does not provide sufficient capacity to support a fiercely competitive packaging industry (Instate, 2000). Problems of price competitiveness within the Australian-New Zealand processed-food sector are compounded by a series of nonprice factors that impinge on the ability to capture export markets. There is a low level of research and development expenditure in the Australian processed food sector (Australian Bureau of Statistics, 2000) and, with the notable exception of wine, industry analysts have bemoaned an apparent absence of an "innovation culture" within the sector. Furthermore, in some industries vertical integration effected by foreign-owned firms may limit value-adding opportunities domestically. In the beef industry for instance, it has been suggested that foreign interests may be inclined to adopt pricing and marketing practices that marginalize antipodean interests (Australia: Queensland Government, 1997).

An array of firm-specific factors interact with the industry-specific factors, described previously, to create the internationalization paths of antipodean agrofood firms. To explore these issues, the remainder of this chapter is devoted to three case studies of internationalization. A case study approach reveals the nuances and vagaries of firm behavior, and thus provides the optimal means to highlight the ways internationalization needs to be understood in terms of an intersection between industry- and firm-specific variables. Because the case studies selected here are the three most important examples of antipodean food MNE internationalization, they contain the key narratives of firm behavior relevant to the issue at hand.

AUSTRALIA/NEW ZEALAND AS A BASE FOR GLOBAL BRANDED FOOD COMPANIES: THE FOSTER'S GROUP

During the 1990s the Foster's Group (FG), known as the Foster's Brewing Group (FBG) prior to July 2001, employed the most successful international expansion strategy of any antipodean food firm. This success rested centrally on its capacity to develop and exploit key intellectual property—in particular, the Foster's beer trademark. Ownership-specific advantages were pivotal to FG's internationalization in the 1990s. FG leveraged these O advantages by restructuring its productive assets and activities over the decade. Whereas in the 1980s the company used the acquisition of breweries in offshore nations to expand its brands, in the 1990s these interests were divested in favor of production-by-licensing agreements. The evolution of this strategy has seen the focus of FG's internationalization shift from the management of tangible production to the management of intangible brand assets. With this, the company's I advantages have undergone transition, as the logic of FG's international activities now are tied more closely to brand development and stewardship. In tandem with these developments, however, FG's range of interests has also shifted, as the company has diversified into wine and leisure businesses.

FG has particular interest for this chapter, because it not only developed the most successful internationalization strategy of any antipodean food firm, but also focuses attention to the question of whether, in a global economy, it is possible to develop and manage internationally branded products from (the relatively small and distant) territories of Australia and New Zealand. The FG experience provides a cautious yes to this question. At the same time, however, emerging consolidation in the global brewing sector may shift the basis for competitiveness in these activities, and thus may require FG to adopt new internationalization strategies. In short, the successful strategies of the recent past may not necessarily provide a blueprint for the future.

The successes of FG over the past decade would have seemed an unlikely outcome in the early 1990s. The forerunner to FG, Carlton and United Breweries (CUB), grew rapidly in the 1970s and 1980s through consolidation of the domestic Australian beer market. By the mid-1980s, two corporate groups (CUB and Tooheys-Castlemaine, now known as Lion Nathan) had captured over 85 percent of Australian beer sales. This consolidation occurred in the context of great corporate volatility. For most of the 1980s, CUB was a division of Elders IXL Ltd, a highly diversified conglomerate (Fagan, 1990). A hostile takeover bid in 1985 destabilized the company's stock registry and then, in 1989-1990, the company was subjected to a contested management buyout by the company's chief executive officer, John Elliott. In the wake of complex financial transactions relating to the attempted buyout, the Japanese brewer Asahi purchased 17 percent of the company, Elliott's plans were scuttled, and FG was effectively bankrupted.

Of course, Foster's was not alone in the 1980s in facing great instability in its stock registry. The Anglo-American merger and acquisition wave during this era had particular focus on branded food companies (Pritchard, 1994). Nevertheless, FBG entered the 1990s with a relatively weak financial position and an unwieldy equity base. FBG surmounted these problems through the successful articulation of a two-pronged internationalization strategy through the decade. First, the company focused its international beer operations on the intangible asset value of the Foster's brand, rather than the attempt to increase its own production of beer. Second, FBG addressed falling per capita beer consumption in core Anglo-American markets through strategic diversifications into wine.

Strengthening the emphasis on the Foster's brand during the 1990s required a dramatic shift in corporate strategy. Under the corporate leadership of Elliott, FBG had acquired substantial beer-brewing operations in the United Kingdom (Courage) and Canada (Molsons). In the 1990s these were divested. In the United Kingdom, Courage had been a poor performer, with FBG having to reduce its investment book value by AU$1 billion in the early 1990s. In 1995 FBG sold the business to the U.K. brewing company Scottish and Newcastle Plc. As part of the Courage divestment, FBG entered into a licensing agreement with the company's new owners for the sales and marketing of Foster's beer in the United Kingdom. This deal was reported at the time to guarantee FBG a "minimum royalty payment of AU$152 million over ten years" (Kaye, 1995, p. 25). By the end of the 1990s, this strategy had evolved to such an extent that Foster's had become Scottish and Newcastle's largest-selling brand. In 1998, the company divested the Canadian Molson business, and under broadly similar arrangements, operated the North American rights to the Foster's brand through a three-way joint venture partnership with Molson USA and Miller (however, Molson left the partnership in 2000). These transactions demonstrate the central and strategic role of brand management within FBG. By divesting Courage and Molson, FBG effectively exited the U.K. and Canadian beer manufacturing and hotel industries, in favor of an outsourcing strategy whereby third parties would manufacture Foster's beer under licence. As FBG's president said in 1998, "There will be rising profit contribution from the brand" (AAP Information Services, 1998). Reflecting the importance of brand management, in March 2000 FBG announced it would establish a new division, Foster's International, with the sole function of brand stewardship.

The attempt to build the Foster's brand globally saw efforts to introduce the beer into emerging market economies. The most significant of these was China, where the spectre of a potentially immense market beguiled company management. The confluence of rapid economic growth and shifts in consumer cultures saw Chinese beer demand grow by 20 percent annually over the 1980s and 1990s (AAP Information Services, 1998). FBG entered the Chinese market during 1992-1995 with the purchase of three breweries. By 1998, however, the group had recognized problems in its "overaggressive entry into China," reducing the book value of its Chinese investment by AU$168 million and closing two breweries (AAP Information Services, 1998). These failures seem to have hinged on intense competition within the Chinese beer market: when FBG was plotting its China strategy, so was almost every international beer company, leading to a crowded marketplace. Unlike some of its competitors, which pulled out of China in the late 1990s, FBG stayed in the market with a single brewery in Shanghai and, late in the decade, established businesses in India and Vietnam to offset its lower-than-expected sales performance in China.

FG's international promotion of the Foster's brand through licensing agreements needs to be contextualized in terms of potential global concentration of the brewing industry. The world's four largest beer companies account only for 20 percent of the world market and, according to some industry analysts, this provides conditions facilitating mergers (by comparison, the four largest soft drink manufacturers control 80 percent of their global market [Benson-Armer, Leibowitz, and Ramachandran, 1999, p. 111]). Twice in recent history (1996 and 2001), market rumors have suggested an impending takeover offer for FBG by Heineken, and in 2000 there were rumors that Diageo, the U.K. company that owns Guinness, among other brands, would make a takeover offer. In these circumstances, it is not unreasonable to speculate that the O advantages of brand stewardship may be served equally well through the incorporation of the Foster's brand within a corporate entity with a broader international brand portfolio. In any such eventuality, individual FG stockholders may benefit, but the Foster's brand itself would be detached from its Australian corporate parentage and national roots.

Regardless of what the future for the Foster's brand may hold, in any case, FG has diversified in recent years, reducing its reliance on beer. One of the key ironies of FBG's recent corporate experiences is that, at the same time the company was attempting to establish Foster's as a global brand, demand in its core markets was static. Through the 1990s FBG remained a successful and profitable operator in the Australian beer market. The company possessed 50 to 55 percent market share in a relatively stable duopoly. However, per capita Australian beer consumption fell 30 percent from the mid-1970s to the early 1990s. Faced with this market environment, during the

1990s FBG extended its alcohol interests to wine, spirits, and linked leisure businesses. The centrepiece of this shift was the 1995 acquisition of Mildara Blass, Australia's fourth-largest wine company. Subsequent acquisitions of smaller Australian wineries and strategic winery purchases in California, France, and Chile boosted the Mildara Blass business through the 1990s until, in October 2000, the company announced a massive AU$2.6 billion purchase of California's Beringer Wine Estates. This acquisition doubled FBG's wine sector turnover, propelling it to become the third-largest wine company in the world. Prior to the Beringer acquisition Australian sales contributed 54 percent of Mildara Blass's turnover; afterward, 66 percent of the combined group's turnover was generated in North America (AAP Information Service, 2001). Symptomatic of this new geographical scope, in December 2000 the renamed wine business Beringer Blass purchased the Castello di Gabbiano winery in Italy, which manufactures the second most popular Italian Chianti imported into the United States.

The expansion into wine was complemented by strategic acquisitions in wine marketing, spirits, and leisure industries. In 1997, FBG purchased Cellarmaster Wines, at the time the world's second-largest mail-order wine business, representing 5 percent of Australian bottled wine sales (Jemison, 1997). With this purchase, FBG also acquired Australia's largest contract wine-bottling operation. In 1999-2000 the company acquired additional mail-order wine businesses in Australia, Germany, California, the Netherlands, and Japan, and claimed to control a 10 percent world share of this sector (Mitchell, 2000). These marketing and distribution businesses were well suited to the emergence of "B2C" e-commerce, and in 2001 FBG commanded two of the three largest Australian online wine Internet sites (Bryant, 2001). The purchase of mail-order and e-commerce wine businesses facilitated FBG's expansion into the distribution component of the wine value chain, potentially enhancing the profile and position of Mildara Blass wines in market channels. During this same period FBG entered the spirits industry with the purchase of Seagram's Australian and New Zealand operations and executed a series of commercial deals that resulted in owning 159 hotels, 80 bottle shops (retail alcohol stores), and the rights over 800 gaming (slot) machines in Australia. Also in 2001, FBG purchased Australia's largest wine barrel manufacturer, providing an additional element of vertical integration to its supply chain arrangements. The significance of this expansion into wine and related businesses was underlined when, in 2001, the Foster's Brewing Group was renamed the Foster's Group. In 2002, for the first time, FG's wine and related business revenues exceeded its beer revenues.

FG's internationalization path during the 1990s reflects management's perceptions of the entity's OLI advantages (see Table 7.2). In brewing, brand stewardship found favor over direct production, as the central element of the company's O advantage. In turn, this executed shifts to the firm's L and I advantages, with marketing being internationalized alongside extensive use of strategic alliances and joint ventures. Divestment of brewing assets freed capital for diversification, notably into wine. At the time of writing, FG is a corporate entity with strong, internationally oriented brewing and wine businesses. Yet as the global profit environments for brewing and wine continue to evolve, it is likely that FG will be required to adopt new internationalization strategies.

USING AUSTRALIA/NEW ZEALAND AS A REGIONAL BASE: COCA-COLA AMATIL

Coca-Cola Amatil (CCA) provides a very different case study from FG. First, the company's key O advantage has rested not with its ownership of intellectual property—as discussed in the following, CCA is a bottling firm that uses intellectual property under licence—

TABLE 7.2. The internationalization strategy of the Foster's Group.

	Ownership advantages (the "why" of MNE activity)	Location advantages (the "where" of production)	Internalization advantages (the "how" of involvement)
Brewing	Ownership of intangible assets (brands and recipes) provides the basis for market expansion	Production (under licence) close to markets in United Kingdom, United States, etc., to reduce transport costs of finished products	Production under licence allows control of brand and recipe assets, creating internal royalty stream
Wine	Capture of California wine production as complementary assets	Production close to raw materials (wine grapes)	Takeover of existing production to build global scale

but with alleged management expertise that has allowed it to purchase market franchises for globally branded products (Coca-Cola and associated soft drinks). Second, in terms of location-specific factors, CCA's internationalization over the past decade has focused on newly emerging markets in Eastern Europe and the Asia-Pacific. Finally, unlike FG, which has been successful in its internationalization, CCA activities have been characterized by reverses in strategy and significant financial losses.

For these reasons, CCA provides a useful counterpoint to FG. In particular, it focuses attention to the question of whether Australia and New Zealand's relative geographical proximity to Asia-Pacific markets, in the context of potential Asia-Pacific trade liberalization and the possibilities of dietary changes in Asia toward Western foods, provides a basis for antipodean food firm internationalization. Certainly over the past decade governments in both countries have enthusiastically embraced the notion that Asian markets provide fertile ground for investments by antipodean food firms (Pritchard, 1999a). Recent history, however, brings to attention the complexities of such assumptions. Australia and New Zealand have extensive Asian agrofood trading links, but historically these have consisted mainly of primary commodity exports. Rapid economic growth in Asia during the 1980s and first half of the 1990s raised the prospect that this "shallow integration" (UNCTAD, 1994) could lead to the "deeper integration" of investment. Yet as this case study illustrates, investment linkages between Australia-New Zealand and Asian nations are positioned within wider, global-scale capital networks, which impinge upon their shape and success. These wider networks were especially pertinent in CCA's case, because of equity and managerial linkages between CCA and the global Coca-Cola system.

CCA's preeminent business activity is to bottle and market Coca-Cola products. As such, its existence remains within the locus of The Coca-Cola Company (TCCC) of Atlanta, Georgia. Since the 1920s, the corporate structure of the Coca-Cola system has rested on a division of responsibilities whereby TCCC owns the relevant trademarks and manufactures the concentrate (in accordance with the famous "secret" recipe), while bottling and distribution is undertaken under agreement by bottling firms. From TCCC's perspective, this structure serves to focus the company's efforts on intellectual property management, with manufacturing and distribution functions outsourced.

TCCC sells bottling firms the concentrate under established terms and conditions (estimated at approximately 20 percent of sales revenue [Mitchell, 2001a]). To enhance its control of the system, TCCC often takes an equity stake in bottling firms. In CCA's case, TCCC purchased a majority stake in the company in 1989, which it held until 1995 after which it reduced its equity progressively to about 35 percent.

CCA became a significant bottling firm in the late 1980s. Prior to 1990, CCA's main business activity was tobacco. Declining cigarette consumption combined with the threat of lawsuits encouraged the company to diversify into the soft drinks sector in the late 1980s and finally, in 1990, it sold its tobacco interests. During the following years CCA consolidated its Coca-Cola businesses into a single operation, and by 1993 it controlled all but one Coca-Cola franchise in Australia, as well as franchises for New Zealand, Papua New Guinea, and Fiji. These operations provided a steady and reliable income stream. In the early 1990s Coca-Cola and Diet Coke together accounted for over half of all soft drinks sold in Australia. When other CCA brands were added (including Fanta and Sprite) the company's market share exceeded 66 percent (Shoebridge, 1995).

In the early 1990s, CCA's position as a sizeable bottling firm focusing exclusively on Coca-Cola products coincided with desires by TCCC to rationalize the number of bottling firms and create "anchor bottlers" responsible for large market territories. With the imprimatur of TCCC, this created an opportunity for CCA to aggressively expand its ownership of Coca-Cola franchises. In so doing, CCA targeted the world's two fastest-growing regional markets for Coca-Cola; the post-Communist states of Eastern Europe and the rapidly growing economies of East and Southeast Asia. By 1995, CCA owned 100 percent of the Coca-Cola market in Hungary, the Czech Republic, Slovakia, Belarus, Ukraine, Romania, and Slovenia; 85 percent in Austria; 75 percent in Croatia; and 60 percent in Poland. In Asia, CCA began acquiring local franchises in Indonesia in 1993 and within two years had captured all franchises in the country and commenced work on the world's largest bottling factory, in Jakarta. During 1995-1996, Indonesian Coca-Cola sales increased by 25 percent per annum (Ferguson, 1996).

Two deals in 1996-1997 catapulted CCA's Asian strategy beyond Indonesia. In August 1996 the Kerry Group, based in Malaysia and owned by entrepreneur Robert Kuok, negotiated to buy 9.2 percent of the company's stock. This investment had strategic importance because the Kerry Group owned extensive Coca-Cola franchises in China and was a key supplier of sugar (a major ingredient in soft drinks) throughout the Asian region. Immediately following the formation of this strategic alliance, CCA's sugar contracts were renegotiated. Furthermore, the Kerry Group's stock purchase generated AU$667 million in cash for CCA, which could be invested for further expansion. Then, in April 1997, CCA announced a AU$3.4 billion merger with the San Miguel Corporation of the Philippines. The deal involved CCA taking control of the San Miguel–owned Coca-Cola Bottlers Philippines Inc., in exchange for a 25 percent share of CCA equity and three seats on the CCA board. Through these acquisitions, in the space of a few years CCA became the second-largest Coca-Cola bottling firm in the world, and the largest outside the United States.

On the basis of these deals, CCA's Australian management painted the company as an "Australian success story." For a period, investors seemed to agree. Strong positive sentiment over CCA's strategy and investments allowed the company's stock to trade at an extremely high price to earnings ratio; by October 1996 its stock price was at fifty-eight times earnings (Ferguson, 1996). This success, however, was built on an unresolved tension within CCA. Although in 1995 TCCC reduced its stake in CCA to less than 50 percent, CCA was never a wholly independent entity. At the 1996 annual general meeting some independent stockholders questioned the influence of TCCC in determining CCA's corporate strategy. To that point, CCA's offshore expansion strategy had rested squarely on maximizing sales rather than building profit margins or optimizing dividend payouts. CCA management defended this strategy on the grounds that the capture of franchises in the short term would generate long-term profits. This was not an unreasonable argument, but, as some analysts also pointed out, the major beneficiary of the sales-maximization strategy was TCCC, because increased Coca-Cola sales translated directly to increased demand for Coca-Cola concentrate. Prima facie there appeared the possibility of a conflict of interest between the aspirations of TCCC (approximately 35 percent of CCA stock), which looked to sales growth as the key measure of performance, and other stockholders, to whom profits and dividends were the cornerstones of corporate performance.

These tensions were exposed in November 1996 when, following a lower-than-expected quarterly profit result, the CCA stock price fell from AU$18 to AU$14.15 (Knight, 1996). The San Miguel merger five months later further exposed the depth of these tensions. Although the price of CCA's stock rose in the few months following the merger, by July 1997 it began to slide again. These fears ballooned two months later, in September 1997, with the onset of the Asian financial crisis. The Philippine peso fell by 25 percent against the Australian dollar in the first few weeks of the crisis and, because CCA's Filipino earnings were not hedged, this cut profits by AU$70 million, or 20 percent (Deans, 1997). CCA's Indonesian earnings also evaporated with the collapse of the Indonesian rupiah and the onset of political crisis. In February 1998 CCA attempted to extricate itself from these problems when it announced a controversial AU$2.3 billion restructure. This involved the purchase of South Korean Coca-Cola franchises for AU$877 million (from TCCC) and the demerger of the company's European operations. CCA's European businesses, along with a number of Italian Coca-Cola franchises, were packaged into a new, London-listed company called Coca-Cola Beverages Plc. Through a stock swap, CCA stockholders were provided with equity in this new company.

The restructure was controversial because it signaled an abrupt change of direction for CCA. After years of developing its European businesses, these were unceremoniously sold. As an influential Australian business analyst commented,

> The divestment of Europe[an businesses by CCA] is a unique step for a major public company. At the end of the process, CCA will have divested itself of all its European operations without getting a cent for them. Any benefits will flow instead to its shareholders. (Sykes, 1998, p. 21)

According to CCA management, European franchises were sold because of falling profitability. Although CCA's European profits fell during 1996, some analysts pointed out that this decline could be attributed to an increase in the price of Coca-Cola concentrate charged by TCCC to its European bottling firms (Mitchell, 1998). If this is correct (TCCC jealously guards such information) it amounts to a situation in which TCCC boosted profits (through raising concentrate prices) at CCA's expense. Moreover, the sale of Eastern European franchises occurred at the same time that CCA was buying South Korean franchises from TCCC, relieving the parent of these relatively high-risk businesses (Mitchell, 1998). Some institutional investors seemed to have interpreted these events as implying CCA's strategies were being vetted by TCCC, which added to the weakening of the stock price over this period.

In any case, prolonged economic crisis in Indonesia and poor performances in South Korea and the Philippines amplified CCA's difficulties. As the end of the 1990s grew near, San Miguel put up for sale its 21.5 percent stake in CCA but, finding little interest, later withdrew the offer. In early 2000, CCA reduced the book value of its Asian assets and, in February 2001, consummated a deal whereby San Miguel would sell its CCA stake in exchange for gaining control (again) of the Philippine Coca-Cola franchise. CCA's unglamorous exit from the Philippine market saw the company post a loss of AU$1.1 billion from its initial investment (Mitchell, 2001b) and reinvented the company as a smaller, more Australian-focused, operation.

CCA's troubled forays into Eastern Europe and Asia were salvaged only by rapid growth of soft drink consumption within Australia. Between 1988 and 1998 Australian per capita soft drink consumption rose 32.5 percent (Shoebridge, 1999). In the late 1990s CCA's earnings were further boosted by 30 percent average annual growth in the bottled water market, within which CCA held the leading brand (Mount Franklin, with 30 percent market share). A disease outbreak in Sydney's water supply during July-August 1998 caused an overnight tripling of bottled water sales and a long-term boost in bottled water consumption. Growth in demand has been critical to CCA, because with 65 percent market share, anticompetitive concerns effectively prevent growth by acquisition in the domestic market. In 1999, Australia's competition regulator rejected CCA's proposed acquisition of Schweppes' Australian soft drink brands following a global AU$2.9 billion deal between TCCC and Cadbury Schweppes. The acquisition of the Schweppes' Australian brands would have allowed CCA's market share to climb from approximately 66 percent to 80 percent. A proposal by CCA to cherry-pick Schweppes' most widely known international brands (Dr Pepper and Canada Dry) and divest all others was rejected, and in the following year Cadbury Schweppes' Australian operations were merged with CCA's key rival, Pepsi-Cola Bottlers Australia, creating a domestic duopoly.

For CCA, the 1990s was a decade of grand dreams and abrupt shifts in corporate strategy. The company failed in its attempt to become an anchor bottling firm for the Asia-Pacific region, bolstered by an exposure to high growth markets in Eastern Europe. By 2002, it was a smaller entity with a cash cow business in the domestic Australian market and a few offshore interests (Indonesia and South Korea). At one level, the CCA experience highlights the difficulties for new entrants in "doing business in Asia" (Pritchard and Lloyd, 2003). At another level, however, it highlights how antipodean-Asian linkages do not exist in isolation, but comprise a part of wider capital networks. These factors highlight the limitations of CCA's assumed OLI advantages during the 1990s (see Box 7.1). Poor profitability in Asia exposed the brittleness of management claims that CCA possessed special skills (O advantages) in running bottling franchises. Location-specific (L) factors, in terms of CCA's participation in fast-growing newly emerging markets, were exposed as being liabilities, with the Asian financial crisis of 1997-1998. Moreover, CCA's assumed I advantages—its privileged status as an anchor bottler within the Coca-Cola system—were found wanting by the end of the decade. CCA's internationalization path during the 1990s provides a cautionary tale illuminating the limitations of Australia and New Zealand as potential home nations for food MNEs in the Asia-Pacific.

LEVERAGING COMPETITIVE ADVANTAGE: THE FONTERRA STORY

The third case study is the New Zealand–headquartered dairy cooperative Fonterra. In contrast to the previous examples, the Fonterra case reveals an internationalization strategy based on profit capture from internationally competitive agrocommodity production. Accordingly, in possessing O advantages relating to raw material supply, the Fonterra case best reflects the argument, raised at the outset of this chapter, that the competitive agricultures of Australia and New Zealand may pave the way for internationalization by local firms. However, and this is the key component of the Fonterra narrative, although the international competitiveness of New Zealand dairy pro-

---

BOX 7.1. The Internationalization Strategy of Coca-Cola Amatil

Ownership advantages (the "why" of MNE activity)	Location advantages (the "where" of production)	Internalization advantages (the "how" of involvement)
Managerial skills in bottling provided the justification	Bottling needs to be undertaken close to customers, in overseas locations	Strategic links with The Coca-Cola Company as an authorized "anchor bottling firm"

---

duction provides the platform for Fonterra's international expansion, translating this competitive advantage into the basis for a viable antipodean-based MNE has required supportive government actions. As this case study illustrates, Fonterra's emergence is tied to the conjoined effects of agrocommodity competitiveness and the specific conditions of the institutional/regulatory environment.

A New Zealand Act of Parliament established Fonterra in October 2001. Upon formation, Fonterra comprised the former operations of the New Zealand Dairy Board, the statutory single-desk export agency for New Zealand dairy products, plus the activities of New Zealand's two largest dairy cooperatives, Kiwi Cooperative Dairies (Kiwi) and the New Zealand Dairy Group (NZDG). The merger of these three entities created a cooperatively owned business with annual turnover of approximately US$5 billion, representing 20 percent of New Zealand exports and 7 percent of the nation's GDP. At the time of its establishment, Fonterra was the world's ninth-largest dairy business, with 120 subsidiaries in forty countries. Seventy-five percent of Fonterra's revenue is generated through sales outside of Australia and New Zealand. The majority of this represents the export of processed milk products from New Zealand, although the cooperative also undertakes value adding in host countries.

The creation of this industry giant involved a deft compromise of deregulatory and re-regulatory objectives within the New Zealand government. In line with Cairns Group philosophies of agricultural deregulation, the creation of Fonterra abolished statutory single-desk export monopoly arrangements over New Zealand dairy products, paving the way for free market processes. Yet at the same time, the Act of Parliament establishing Fonterra also represents a deliberate intervention to shape the industry's future structure. The Fonterra legislation allowed the establishment of a single entity with 95 percent control over the New Zealand dairy sector, despite obvious prima facie grounds that this outcome contravened market concentration provisions of the New Zealand Commerce Act. Given New Zealand's recent history as a model for neoliberal experimentation within the global economy (Kelsey, 1995), the re-regulatory implications of the Fonterra legislation provoked considerable debate among some members of the nation's business establishment. The New Zealand (Labor) government's decision to deregulate statutory single-desk exporting and simultaneously intervene in postderegulatory industry outcomes, represents a historical juncture in New Zealand agriculture, whereby the national interest implications of "hand's off" liberalization were scrutinized more intensely, and industry participants were given scope to determine their preferred future industry arrangements.

Evidently, an appreciation of Fonterra's current structure and market position requires consideration of the regulatory arenas from which it has emerged. On both sides of the Tasman Sea (that is, in both Australia and New Zealand) the dairy sector has, until recently, been subject to considerable and complex economic regulation. Of course the dairy industry is highly regulated in many countries, and to this extent the trans-Tasman situation has parallels with experiences elsewhere. However, what distinguishes the Australian and New Zealand dairy sectors is that during the 1990s the respective governments of both countries set about to liberalize dairy sectors and, as a result, industry players underwent significant restructuring.

In each country these processes took different shapes. In New Zealand, the merger of cooperatives provided the key element of dairy restructuring. With 93 percent of New Zealand dairy exported, restructuring processes pivoted on the pricing and purchasing arrangements of the Dairy Board as a monopoly buyer. As Curtis (2001) recounts, the Dairy Board purchased milk products from the dairy cooperatives on the basis of two payment components: a payment for the value of milk fat and protein supplied, and a payment for the estimated manufacturing costs of dairy products (which obviously differed for separate products such as milk powders, butter, and cheese). The Dairy Board estimated manufacturing costs on the basis of a national model assuming an "average-sized" factory. If dairy cooperatives could merge operations into larger and more efficient factories, they could receive above-normal profits from the sale of products to the Dairy

Board. This generated strong encouragement for mergers, leading to the formation of two cooperatives (Kiwi and the NZDG) responsible for over 90 percent of sector output.

Because of its larger domestic market and the legacy of decentralized (state government) powers over food regulation, Australia's regulatory regime was more complex than New Zealand's. Until the late 1980s, Australian dairy regulation comprised minimum farm-gate pricing for fluid milk (called "market milk" in Australia), restrictions on interstate milk trade, and export payments to manufacturing milk producers. The manufacturing milk sector was deregulated with the 1986 Kerin Plan, whereas the market milk sector was deregulated progressively through the 1990s. The last vestige of economic regulation, minimum farm-gate pricing, was abolished in 2000. Pritchard (1996) discusses the complexity of regulatory arrangements and how they intersected with evolving corporate structures. For the purposes of this chapter, the key point is that progressive deregulation of the sector encouraged a series of mergers that built discrete specializations among industry participants. By 2001, the Australian dairy sector had evolved into a bifurcated industry structure of specialist fluid milk and branded commodity producers (National Foods, Dairy Farmers, Parmalat, and Peters and Browne) and specialist manufacturing milk producers (Bonlac and Murray Goulburn).

The progressive deregulation of the Australian industry provided fertile investment opportunities for emerging, large-scale New Zealand dairy cooperatives. By 2001, New Zealand interests had acquired three strategic investments in the Australian dairy sector: the New Zealand Dairy Board possessed a 25 percent interest in Bonlac, the NZDG owned an 18 percent interest in National Foods, and Kiwi owned an 80 percent interest in Peters and Browne. These cross-investments facilitated coordination and integration of the two countries' dairy sectors.

The New Zealand cooperative mergers, deregulation of the Australian industry, and cross investments between the two countries provided the context through which in 2001 the New Zealand government agreed to deregulate the Dairy Board and allow its merger with the NZDG and Kiwi, thus forming Fonterra. This decision established a merged entity with a range of interests in both dairy commodity exports and branded products. To manage these interests, Fonterra created two divisions: New Zealand Milk Products

(NZMP), which operates the group's dairy ingredients business (mainly comprising dairy agrocommodity exports such as milk powders and unbranded butter and cheeses), and New Zealand Milk (NZM), which operates the group's consumer branded products operations. At the time of writing the operational spheres and scope of the two divisions remain in flux. It is probable that as Fonterra evolves, new divisional structures may emerge. Nevertheless, this internal architecture appears to reflect a key element in Fonterra's strategic vision, namely, its ambitions as both a global-leading ingredients supplier and an entity with interests in a range of branded dairy products that build on New Zealand's reputation for quality, environmentally clean foods (see Box 7.2).

The NZM business revolves around a range of products with brands strongly associated with New Zealand, such as Anchor, Fernleaf, and Mainland. NZM dominates the New Zealand domestic consumer dairy products market, although under provisions of the 2001 legislation Fonterra has been forced to sell some interests in fluid milk. In 2002, NZM's Australian operations were merged into a holding company (Australiasian Food Holdings Ltd) and the company acquired an additional 25 percent of Bonlac Foods Ltd, bringing its equity position to 50 percent. Notwithstanding the value of these operations, Fonterra's dairy ingredients business, operated through the NZMP division, represents the company's key area for growth. NZMP has offices in over 100 locations globally, in all inhabited continents. As dairy demand grows globally, in an environment of potentially liberalized trade, NZMP is in a prime position to capture growth and profitability. To execute this potential, Fonterra has initiated a series of strategic alliances in key markets. To date, the largest of these has involved an alliance with Nestle in the Americas. Through a 50:50 partnership, Nestle and Fonterra have agreed to act jointly with

---

BOX 7.2. The Internationalization Strategy of Fonterra

Ownership advantages (the "why" of MNE activity)	Location advantages (the "where" of production)	Internalization advantages (the "how" of involvement)
Securing access for raw materials and exports	Enables local branding and value-adding, averts protective barriers	Takeovers and joint ventures with firms with marketing assets

respect to dairy market development. This is a key development at the global scale, because Latin America is the world's fastest-growing dairy market and, as a result of various regional trade agreements, there is considerable scope for industry restructuring. The Nestle-Fonterra partnership makes use of Nestle's expertise in marketing and branding along with Fonterra's technical knowledge of dairy production, its branch office structures, and the use of Fonterra's New Zealand production as a preferred source of imports. Significantly, because the deal is limited to the Americas it does not restrict Fonterra's scope to compete with Nestle in other regional markets. Thus, the partnership enables Fonterra:

> to co-operate with Nestle in one part of the world; to supply ingredients to Nestle in some parts of the world; and to compete with Nestle with branded products in other parts. [According to Fonterra board member Warren Larsen,] "You don't necessarily have one deal that's global. Instead, you have deals pertinent to a part of the world, driven off the market construction in those areas." (Edlin, 2001: 4)

In Europe, Fonterra has established an alliance with Arla Foods, the continent's largest dairy producer, to cooperate in the large and lucrative U.K. butter market. The agreement involves Arla packing Fonterra's Anchor butter under license. In India, Fonterra has established an alliance with Britannia Foods (part of the Danone Group) for milk and food service. In the United States, Fonterra has established a joint venture with Dairy Farmers of America, the world's largest dairy cooperative, for the production and supply of specialist dairy products. In early 2002, Fonterra management indicated they were eyeing a significant number of potential acquisition targets across the globe, with an intention to double the cooperative's revenue within a decade.

Fonterra's aggressive pursuit of alliances and its lofty growth ambitions raise several issues. First, they presuppose further significant growth in New Zealand dairy production. The dairy industry has grown rapidly in New Zealand during the past decade, particularly in the South Island. With expansion, changes have come to dairy farming. The economies of scale advantages of increased herd sizes dovetails with new ownership structures in dairy farming, opening the way for corporate management and control of dairy farms, within an overall environment of increased capital intensity. Second, Fonterra's international growth raises issues with respect to its status as a supplier-owned cooperative. Dairy farmers own Fonterra equity relative to their supply volumes, thus sharing directly in the cooperative's profits and losses. Accordingly, Fonterra's cooperative status ensures a mechanism whereby benefits from international expansion are transmitted to farmers, embedding its fortunes within the New Zealand - rural economy. Fonterra's 50 percent interest in the Australian farmer-owned dairy producer Bonlac, moreover, provides a basis for the possible creation of a trans-Tasman (New Zealand-Australia) dairy cooperative. Yet at the same time, under the legislation establishing Fonterra, New Zealand farmer-suppliers are empowered to sell up to 20 percent of their milk to competing dairy companies. This opens the way for potential competition to Fonterra. It remains a possibility that despite the successes of Fonterra's international alliances it could face intensified competition in New Zealand for raw milk supply.

Fonterra therefore remains a contradiction. It is the only example of a significantly sized antipodean food MNE whose competitive position has depended on the underlying efficiency of Australia's and New Zealand's agricultural productivity. At the same time, it is an explicit artifact of government regulation. Accordingly, the Fonterra example emphasizes that analysis of food firm internationalization needs to be embedded within an appreciation of the history and political economy of agrofood systems. Food firms are entities rooted in space and time, with configurations of structure and strategy positioned uniquely within landscapes of competition and profit accumulation.

CONCLUSIONS

Australia and New Zealand are important agrofood producers globally, although historically they have not been important home nations for food MNEs. Structural characteristics in these nations' domestic agrofood sectors, especially with respect to agrocommodity export regulation and competition from large, foreign-owned MNEs, have played crucial roles in determining the size and scope of food firm internationalization. Analysis of individual firms reveals wide diversity of internationalization paths within a small number of MNEs. Insights from the work of Dunning emphasize this diversity as an outcome of intersections between industry-specific and firm-specific factors, the latter influenced by articulations of ownership, location, and internalization advantages.

Two clear conclusions are apparent from the material examined. First, there are mixed experiences and uncertain long-term results from the various internationalization strategies pursued by Australian and New Zealand food MNEs since the early 1990s. In the case of FG, the development of global brand recognition has made it an attractive takeover target. The example of CCA provides a cautionary tale over the extent to which Australia and New Zealand's Asia-Pacific location can be used as a launching pad for pan-Asian corporate structures in the food industry. Finally, the Fonterra example emphasises that while Australia's and New Zealand's competitive advantages in agrocommodity production can lead to substantial export-led growth, the embedding of locally controlled corporate structures hinges on specific institutional/regulatory contexts.

Second, evidence suggests a weak relationship between agrofood export competitiveness and the emergence of antipodean MNEs. As mentioned previously, Fonterra sheds uncertain light on this issue, because it is an artifact of regulation. Internationalization of the Australian wine industry may provide some evidence of an association between industry growth and firm internationalization, particularly in the case of FG's acquisition of California's Beringer Estates, but then again, Australia's biggest export wine (Jacob's Creek) is owned by French interests (Orlando Wyndham, a subsidiary of Pernod Ricard). Indeed, other sectors point to the reverse occurring, with the growth of Australia's beef and dairy industries over recent years leading to foreign acquisitions of local interests, as in ConAgra's takeover of Australian Meat Holdings in 1992 and Parmalat's takeover of Paul's (dairy) in 1998. Improved export competitiveness of antipodean agrofood sectors may have the primary effect of creating lucrative targets for foreign MNEs, rather than facilitating local firms to internationalize.

Perhaps these outcomes are not surprising. In a globalizing economy, those countries leading the pace for deregulation necessarily expose their industry sectors to the winds of competition. Whereas prospects for agrofood trade liberalization may encourage an expansion of Australian and New Zealand agroexports, it does not follow necessarily that Australian-New Zealand food MNEs will concurrently emerge.

NOTES

1. *Antipodean,* from the word *antipodes,* is used to describe Australia and New Zealand's status as on the other side of the world's centers of power and influence.
2. As illustrated in Table 7.1, the three case studies represent the three largest antipodean-owned agrofood companies with substantial offshore operations. Goodman Fielder Ltd. was not selected for analysis, because its non-Australian and non-New Zealand interests amounted to just 21 percent of revenue.
3. The one exception to this is Fonterra, which at the time of writing has not published its first annual report. Personal communication with Fonterra management confirms that the only geographical segment information currently available relates to where final sales are made, rather than where value is added.

REFERENCES

Aharoni, Yair (1966). *The foreign investment decision process.* Boston, MA: Harvard University Press.

Anonymous (2001). Deal a meal: The global 150 companies report. *Prepared Foods,* 170(7): 10-16.

Australia: Horticultural Task Force (1994). *Australian horticulture.* Canberra: Australian Government Publishing Service.

Australia: Industry Commission (1991). *Australian dairy industry,* Report 14. Canberra: Australian Government Publishing Service.

Australia: Productivity Commission (1998). *Benchmarking the Australian waterfront.* Canberra: Ausinfo.

Australia: Queensland Government (1997). Submission to Australia: House of Representatives Standing Committee on Rural and Regional Affairs inquiry into Australian trade liberalization. Unpublished. Available from the Department of the House of Representatives, Canberra.

Australian Associated Press (AAP) Information Services (1998). Foster's, admits to taking over-aggressive tack with China. AAP Newswire report, September 28.

Australian Associated Press (AAP) Information Services (2001). Foster's renames wine unit Beringer Blass as it pushes into U.S. AAP Newswire report, February 20.

Australian Bureau of Statistics (2000). Research and experimental development, businesses. Catalogue 8104.0.

Baran, Paul A. and Sweezy, Paul M. (1966). *Monopoly capitalism: An essay on the American economic and social order.* New York: Monthly Review Press.

Benson-Armer, Richard, Leibowitz, Joshua, and Ramachandran, Deepak (1999). Global beer: What's on tap? *McKinsey Quarterly,* 99(1): 111-121.

Bryant, Gayle (2001). Foster's takes a loss on wine planet. *Business Review Weekly (Sydney),* 21(23): 94.

Curtis, Bruce (2001). Forms of governance in the meat and dairy industries of New Zealand. In Lockie, Stewart and Pritchard, Bill (Eds.), *Consuming foods, sustaining environments* (pp. 174-188). Rochhampton, Australia: Central Queensland University and Australian Academic Press.

Dasey, D. (2001). Local fare fails top 100 test of brands. *The Sun-Herald (Sydney),* November 25, p. 29.

Deans, Alan (1997). CC Amatil's attitude problem. *The Australian Financial Review,* October 18, p. 32.

Dicken, Peter (1998). *Global shift.* London: Paul Chapman Publishing.

Dunning, John H. (Ed.) (1971). *The multinational enterprise.* London: Allen and Unwin.

Dunning, John H. (1993). *Multinational enterprises and the global economy.* Reading, MA: Addison-Wesley.

Dunning, John H. (1997). *Alliance capitalism and global business.* New York: Routledge.

Edlin, Robert (2001). Fonterra-Nestle deal puts NZ dairying on top of the world. *Independent Business Weekly (Auckland, New Zealand),* 7(37): 4-5.

Fagan, Robert. (1990). Elders IXL Ltd: Finance capital and the geography of corporate restructuring. *Environment and Planning A,* 22: 647-666.

Ferguson, Adele (1996). Coke's insatiable thirst. *Business Review Weekly (Sydney),* 16(43): 44.

Friedmann, Harriett and McMichael, Philip (1989). Agriculture and the state system: The rise and decline of national agricultures. *Sociologia Ruralis,* 29(2): 93-117.

Goodman, David (1997). World-scale processes and agro-food systems: Critique and research needs. *Review of International Political Economy,* 4(4): 663-687.

Instate (2000). *Exporting Australian processed foods: Are we competitive?* Sydney: Instate Pty Ltd [ISBN 0 642 43253 8].

Jemison, Simon (1997). Foster's Lager: A 160m wine coup. *The Australian Financial Review,* August 5, p. 1.

Kaye, Tony (1995). Foster's sells loss-making British brewer. *The Age (Melbourne),* May 19, p. 25.

Kelsey, Jane (1995). *Economic fundamentalism: The New Zealand experiment—A world model for structural adjustment?* London: Pluto Press.

Knight, Elizabeth (1996). Coca-Cola taxes backroom boffins. *The Sydney Morning Herald,* November 26, p. 29.

Mitchell, Susan (1998). CCA stages war of independence over U.S. parent. *The Australian Financial Review,* March 30, p. 18.

Mitchell, Susan (2000). Foster's eyes overseas labels. *The Australian Financial Review,* March 29, p. 30.

Mitchell, Susan (2001a). Amatil's payments up, marketing support down. *The Australian Financial Review,* March 30, p. 50.

Mitchell, Susan (2001b). Coke takes a $1bn Manila hit. *The Australian Financial Review*, February 7, p. 1.

Platt, D.C.M. and di Tella, Guido (Eds.) (1985). *Argentina, Australia, and Canada: Studies in comparative development*. London: Macmillan.

Pritchard, Bill (1994). Finance capital as an engine of restructuring: The 1980s merger wave. *Journal of Australian Political Economy*, 33: 1-20.

Pritchard, Bill (1999a). Australia as the "supermarket to Asia"? Governance, territory, and political economy in the Australian agri-food system. *Rural Sociology*, 64(2): 284-301.

Pritchard, Bill (1999b). Switzerland's billabong? Brand management in the global food system and Nestlé Australia. In Goss, Jasper, Burch, David, and Lawrence, Geoffrey (Eds.), *Restructuring global and regional agricultures: Transformations in Australasian agri-food economies and spaces* (pp. 23-40). Aldershot, UK: Avebury.

Pritchard, Bill (2000). Geographies of the firm and agro-food corporations in East Asia. *Singapore Journal of Tropical Geography*, 21(3): 246-262.

Pritchard, Bill (2001). Transnationality matters: Related party international transactions in the Australian food industry. *Journal of Australian Political Economy*, 48: 23-45.

Pritchard, Bill and Lloyd, Kate (2003). Business cultures, the state, and the changing investment environment of East and Southeast Asia. In Phelps, Nick and Raines, Philip (Eds.), *The new competition for inward investment* (pp. 173-192). Cheltenham, UK: Edward Elgar.

Pritchard, William [Bill] (1996). Shifts in food regimes and the place of producer co-operatives: Insights from the Australian and United States dairy industries. *Environment and Planning A*, 28: 857-875.

Rama, Ruth (1992). *Investing in food*. Paris: Development Center of the OECD.

Shoebridge, Neil (1995). Study finds Aussies are going to water. *Business Review Weekly (Sydney)*, 15(21): 73.

Shoebridge, Neil (1999). Always seeking sales. *Business Review Weekly (Sydney)*, 19(22): 80.

Sklair, Leslie (1998). Globalization and the corporations: The case of the Californian Fortune Global 500. *International Journal of Urban and Regional Research*, 22(2): 195-216.

Stopford, John. M. and Dunning, John H. (1983). *Multinationals: Company performance and global trends*. London: Macmillan.

Sykes, T. (1998). Keeping an eye on Atlanta. *The Australian Financial Review*, February 7, p. 21.

Symons, Michael (1982). *One continuous picnic: A history of rating in Australia*. Adelaide, Australia: Duck Press.

United Nations Conference on Trade and Development (UNCTAD) (1994). *World investment report*. New York and Geneva: UN Publications.

Chapter 8

Regionalization, Globalization, and Multinational Agribusiness: A Comparative Perspective from Southeast Asia

David Burch
Jasper Goss

INTRODUCTION

Globalization is characterized in much of the literature as a transforming process that has its dynamic element in the activities of the large corporations whose origins are found in the national development project of major industrial countries such as the United States, Japan, or the nations of Western Europe (see, e.g., Dunning and Hamdani, 1997). Such companies came to extend their activities across the globe in a variety of ways; in some cases, this involved direct foreign investment in the establishment of subsidiary companies in other countries, while in other instances, it involved entering into subcontracting arrangements with local entrepreneurs for the production of commodities or services. Whatever the case, the impetus behind such developments was seen to lie with the Western corporations which, in seeking to satisfy the demands of global consumers, determined what was produced, where, and in what quantities.

However, this model of globalization fails to consider the role of those multinational enterprises (MNEs) that have their origins in the less-developed third world. Although some research into third world multinational enterprises (TWMNEs) was carried out in the 1980s (Kumar and Macleod, 1981; Wells, 1983; Lall, 1983; Ting, 1985;

Khan, 1987), this topic remains largely neglected. A number of researchers (Lecraw, 1992; Dunning, van Hoesel, and Narula, 1996; Peng, Au, and Wang, 2001) have revisited this topic from time to time, but there is still little appreciation of the role of third world multinationals and their significance at regional and global levels. Even less research has been directed toward the role of third world agrofood MNEs (see Burch, 1996; Goss, Burch, and Rickson, 1998, 2000; Burch and Goss, 1999; Goss and Burch, 2001; Goss, 2002), despite the fact that a number of major companies have emerged in recent years that not only provide significant theoretical insights into the process of globalization, but also, in some sectors and in numerous sites, challenge the dominant position established by Western multinationals in the agrofood sector.

This chapter attempts to fill some of the gaps in the literature and to offer some insights into the significance of the emergence of multinational agribusinesses from the third world. It seeks to do this by focusing first on some of the recent literature on TWMNEs, which has theorized about the factors behind the push to establish overseas production capabilities. Following from this, we focus on the relationship between this literature and the emergence in recent years of a number of agrofood multinationals from Southeast Asia, utilizing a case study approach which documents the growth of two major companies. We conclude with a comparative analysis that evaluates the significance of these companies and their activities.

APPROACHES TO THE STUDY OF TWMNES

Most of the recent studies of TWMNEs have been conducted by researchers (mostly economists) in the area of business studies and have focused on a variety of issues; some have been concerned with evaluating the contribution third world multinationals might make to national development, while others have considered issues such as technology policy or technological choice. An issue of abiding interest has been the reasons behind the outward push by third world companies. Why would companies located in capital-poor and labor-abundant economies undertake foreign direct investment (FDI) at all? Where did they invest and what benefits did they expect to gain? Was the explanation for this kind of FDI different from the explanation which held for developed country firms, or could some general model explain both?

In their early work on this topic, both Wells (1983) and Lall (1983) focused on the enterprise and argued that TWMNEs differed from developed-country MNEs in a number of ways. In summary, the TWMNEs that emerged in this earlier period tended (among other things) to invest mostly in neighboring less-developed countries, where there were even lower levels of industrial development, in order to reap firm-specific advantages associated with lower production costs, the diversification of risks, the existence of protected markets, and the demand conditions in the host country. They generally adopted smaller-scale and more labor-intensive technologies in the production of goods that were mainly aimed at the domestic market in the host country. The goods produced were usually of lower quality than other MNEs, but they were able to compete effectively on the basis of price rather than quality (Lecraw, 1993).

Another more general approach provided by Dunning (1981, 1988) focused less on the individual enterprise and directed attention instead to explanations for the broad movements in foreign direct investment. Dunning (1981, 1988) suggested that FDI could be explained by reference to the concept of the investment development path (IDP), which suggested that foreign investors experienced three sets of competitive advantage over local companies: ownership advantages, locational advantages, and internalization advantages; these three factors varied in their effects as a consequence of government policy and the level of economic development. Clearly, the focus of the IDP paradigm is broader than the enterprise-specific approach of Wells (1983) and Lall (1983), but it is still country specific. When the model is applied on a country-by-country basis, it reveals a number of inconsistencies and methodological weaknesses. For example, countries are said to experience five main stages of economic development, and the pattern of FDI activity changes as the country in question moves through these stages (Bende-Nabende and Slater, n.d.). However, in their case study of Thailand, Bende-Nabende and Slater (n.d.) suggest that stage 1 of that country's IDP began in 1926 and lasted until 1986 (although the authors are somewhat hazy about the beginning of this stage, they start their narrative from the time when Thailand attained "fiscal autonomy"). More important though, this first stage encompasses a disparate range of political, social, and economic experiences which some observers would find difficult to aggregate in the way the IDP model implies. In addition, despite its apparent acknowledgment of the importance of long-term factors in the process of change and development, the model is essentially ahistorical and does not appear to attach any significance to the factors that shaped Thailand's 1926 starting point. In addition, the weaknesses of the model are especially evident in its application to the agrofood sector since—as will be seen later—at least one major Thai agribusiness company, the Charoen Pokphand (CP) Group, had started its program of FDI nearly a decade before the second stage of policy liberalization and diversified export-oriented manufacturing was said to have begun (Bende-Nabende and Slater, n.d.). Similarly, a "stages of development" model of the kind postulated by Dunning (1981) implies some degree of uniformity in the level of development of sectors involved in FDI. However, it is clear that as far as the food and fiber sector is concerned such uniformity is difficult to find, and the evidence suggests that it is clearly possible to establish a highly sophisticated and productive food-processing sector on the basis of a variety of models of ownership and organization, e.g., peasant farming, plantation production, corporate agriculture, cooperative production, and collective ownership.

In our view, neither the country-specific approach of Dunning nor the enterprise-specific approach of Wells (1983) and others offers an effective framework for a full explanation for the decisions by third world companies in the agrofood sector to undertake FDI activities. Somewhat more insightful is the work by Lecraw (1993) on FDI by Indonesian companies, in which he postulates two kinds of FDI. The first is "export-enhancing" FDI which results from falling trade barriers and the pressures on local companies to become larger, to reduce their costs, and to become more efficient; one response to this situation is for local companies to acquire the technology and management expertise available overseas and access export markets. In general, companies in this category would be relatively small and inefficient in terms of scale and costs, and could be expected to invest mainly in higher-income countries.

Second is "operations-extending" FDI, which involves relatively large, efficient, domestically oriented companies investing overseas in order to serve the domestic market of the investment-receiving country. FDI of this kind would normally be oriented toward other low-income countries where conditions were similar to those prevailing in the home market of the investing company. However, as with the earlier work by Dunning (1981), this approach leaves some important questions unanswered. For example, why do some firms engage in FDI while others in the same sector do not? Lecraw (1993) suggests that the answer to this may partly lie in the lack of managerial expertise on the part of those companies that failed to undertake FDI. However, our work suggests that the situation is more complex than this and that a wider range of explanatory variables is likely to be involved.

It is argued in this chapter that in order to gain a fuller understanding of the dynamics of FDI from less-developed countries, more attention needs to be directed toward a historical and comparative analysis of the companies involved, which considers not only the particular histories and activities of these organizations, but also the commonalities and differences they exhibit. In this chapter, for example, we are analyzing two companies that can be classified as large, diversified conglomerates. The role of the conglomerate company in Asia—the large, multisectoral company exemplified by the *chaebols* of South Korea (such as Hyundai) and the *keiretsus* of Japan (such as Mitsui and Matsushita)—is particularly important in understanding the dynamics of economic growth in the region. But while having much in common, not all so-called Asian conglomerates are the same, and it is important to understand from a historical and a cultural perspective how and why they differ. The two companies selected for this comparative analysis—the Charoen Pokphand (CP) Group of Thailand and the Salim Group of Indonesia—were both originally national companies that emerged from the agrofood sector of their home country. As a consequence, they demonstrate very effectively the dynamics behind the process of transformation which has enabled many other third world agrofood multinationals (such as the San Miguel and Jollibee Corporations in the Philippines, the President Group of Taiwan, and others) to establish a competitive presence in regional and global markets. On the other hand, there are also significant differences between the two companies being studied, which should serve to ensure that we do not over-generalize across all of these various enterprises.

AGROFOOD MULTINATIONAL COMPANIES IN ASIA: UNWIELDY GIANTS OR NIMBLE OPPORTUNISTS?

Many of the countries of East and Southeast Asia have achieved very high levels of economic growth in recent years, largely on the basis of a model of export-oriented industrialization (EOI). According to the World Bank (1994), the experience of the "Tiger" economies—Hong Kong, Singapore, Taiwan, and South Korea—demonstrated that industrialization and development in the third world was possible and that poor rural societies could be transformed into wealthy, urbanized, and industrialized societies operating at the most advanced level of technological development. In addition, this model implied the existence of a second tier of proto-NICs (newly industrializing countries) waiting in the wings, which would replicate this experience. If the correct policies were pursued, there was no reason why Thailand, the Philippines, Malaysia, and Indonesia should not also follow a similar path of development, involving an expanding export-oriented industrial sector and a declining rural sector, measured by diminishing employment in rural areas and a decline in the share of agriculture in GDP and exports.

Many of these illusions were shattered by the Asian economic crisis of 1997, which not only demonstrated the fallacy of elevating particular theoretical models to the status of universal truths, but also reiterated the need to historically situate the experience of particular companies and countries. In contrast to the assumptions made by many market-oriented analysts, the historical and cultural differences among NICs were particularly important when it came to understanding their responses to the events of 1997 and their capacity for postcrisis renewal. There are, of course, many common experiences that the city states of Hong Kong and Singapore shared with the recently agrarian economies of Taiwan and South Korea, but there are also significant differences which have been masked by blanket references to the "Tiger economies" and models of export-oriented industrialization. Similarly, there has been a tendency to characterize the large, diversified conglomerate company as a uniquely Asian form of organization and to place an undue emphasis on those features that such companies are said to share. For example, it is suggested that many such companies have a distinct ethnic base, having been established by overseas Chinese entrepreneurs who were a part of the large prewar and postwar diaspora which saw Chinese capital emerge as the dominant force in the business life of a number of Southeast and East Asian countries. Second, this shared ethnic base is said to have resulted in the establishment of elaborate business networks among individuals and companies, reportedly based on Confucian values of interdependence and a commitment to group identity and welfare, which bring mutual benefits to all companies within the network (Richter, 1999). Third, it is suggested that the Asian conglomerate company relies heavily upon family members to staff senior management positions, rather than employing business professionals whose commitment is secured by high salaries and other benefits. Finally, a close relationship is said to exist between the conglomerate company and the state, which is expressed in a variety of ways, ranging from privileged access to senior decision makers on the one hand, to state patronage—legal or otherwise—in the granting of monopolies, contracts, sinecures, subsidies, tax and concessions on the other (Richter, 1999; Hamlin, 2000).

Such propositions were normally advanced in order to demonstrate that that the Asian conglomerate company was an anachronism that had outlived is usefulness. The unplanned opportunity-driven diversification which was the basis for the growth of many conglomerates had, it was suggested, become a liability by the time the 1997 economic crisis hit the Asia region. With the continuing growth and diversification of the conglomerate company, managers lost sight of their core business as they attempted to exert an ever-widening span of control over the disparate elements of the enterprise. The conglomerate form disguised the operations of the loss-making elements that were draining the profitable parts of the corporation, while the company as a whole had become insulated from the discipline of the market by overdependence on state patronage. For many analysts, the economic crisis of 1997 signaled the end of the conglomerate model (Hamlin, 2000; Hiscock, 2000).

However, the survival and regeneration of many of the so-called Asian conglomerates suggests such propositions are of dubious merit and questionable accuracy. Recent evidence has emerged which suggests that the conglomerate form is not confined to Asian companies. More to the point, the evidence suggests that organizational form is not necessarily the critical determinant of performance or capability, and that large and diversified conglomerate companies can be as efficient and profitable as more focused corporations (Boston Consulting Group, 2002). Some, of course, performed less well, and it is this difference in performance that makes a comparative and historical analysis of third world agrofood multinationals so important. In the following sections, the development of the two companies under review is traced, from their origins within a national agrofood system to their emergence as fully fledged multinational agribusiness companies. Following this, a comparative analysis along a number of dimensions is carried out, which should reveal a number of critical differences (as well as similarities) that explain a company's performance.

THE CHAROEN POKPHAND GROUP

The Charoen Pokphand Group is Thailand's largest agribusiness company. It was established by two migrant Chinese brothers, Chia Ek Chaw and Chia Seow Whooy, who left Shantou in Guangdong province in 1921 and opened the Chia Tai Seeds and Agricultural Company in a small shop house in Bangkok's Chinatown, selling seeds, fertilizers, and insecticides. The company established a feed mill in 1953, which provided the basis for the group's major activities in animal feed and livestock production.

The group's current CEO, Dhanin Chearavanont, a second-generation family member and Chia Ek Chaw's fourth son, took control of the business in 1964 and initiated a period of rapid growth and diversification. By the 1960s, the CP Group was operating two feed mills, and in 1973 it established the Bangkok Farm Company. This was a vertically coordinated poultry operation that contracted with Thai farmers to raise chickens, which were subsequently processed in the company's plant before being shipped to Japan or sold on the local market. The CP Group applied the same model of contract production and vertical coordination as it expanded its Thai-based operations to include other commodities, including pork, fruits and vegetables, seed production, dairying, and, initially, prawn aquaculture. By 1993, the CP Group had emerged as the world's second-largest producer of poultry, the third-largest producer of animal feed, and the largest producer of prawn feed (for a detailed history of the CP Group, see Burch, 1996; Goss, Burch, and Rickson, 1998, 2000; Burch and Goss, 1999; Goss and Burch, 2001; Goss, 2002).

At the same time it was expanding its agrofood operations in Thailand, the CP Group came to be associated with a wide range of agricultural and nonagricultural activities in both Thailand and overseas. Within Thailand, the group invested in the food retail sector, involving supermarkets and hypermarkets, and the operation of a large number of franchise stores, including the Thai franchise for about one-third of the Kentucky Fried Chicken (KFC) outlets established in Thailand. KFC opened its first store in Thailand in 1984; by 2001, at the peak of its operations under the CP Group, it was the leading chain in the Thai fast-food sector, with over 300 outlets and a turnover of 4.5 billion baht (US$53 million). As a consequence of the group's involvement in the fast-food sector, it was able to vertically integrate its operations to the maximum extent and involve itself in the whole supply chain, from the supply of day-old chicks and supporting inputs to the sale of chicken products in its KFC outlets. As a result, the group was able to maximize its returns from the operation as a whole and to minimize the transactions costs associated with those activities that were not fully vertically integrated. Moreover, these benefits were multiplied because of the group's product diversity and its geographical spread; it was not only the main supplier of poultry products to its own and other KFC franchises in Thailand, but also a major supplier of poultry products to other fast-food operators in Thailand, such as Pizza Hut and McDonalds, as well as to the KFC operations in Singapore and other countries in the region (Burch and Goss, 1999).

The CP Group was able to further vertically integrate through its ownership of processing and retailing capabilities when, in 1989, it signed a franchise agreement with the U.S.-based Southland Corporation to establish 120 7-Eleven stores in Thailand. The group's plan for these stores involved the sale of chicken products and other (food and nonfood) lines produced by the group's many subsidiaries. By 2001, the group had opened over 1,500 stores, serving 1.2 million customers daily, and was the world's third-largest operator of 7-Eleven outlets. Seventy percent of store sales were food items, most of which were sourced from within the CP Group itself. Other major areas of investment within Thailand were undertaken in a wide range of nonfood sectors, including fertilizers, pesticides, and agrochemicals, tractors, jute-backed carpets, cement, petrochemicals, vehicle parts, telecommunications, insurance, and property development.

These activities created additional opportunities for increased income streams through horizontal as well as vertical integration across many sectors. For example, Thai consumers were able to purchase a telephone line from the CP-owned TelecomAsia and pay their TelecomAsia telephone bill at any CP-owned 7-Eleven convenience store (Burch, 1996; Burch and Goss, 1999, Goss, Burch and Rickson, 2000; Goss, 2002).

The group also expanded outside Thailand and invested in a host of projects in agriculture, food processing, industry, telecommunications, petroleum marketing, property development, insurance, and much more throughout Asia and the countries of the former Eastern Europe. Its overseas agrofood operations came to include poultry production and processing in Turkey, China, Malaysia, the United States, and Indonesia; animal feed operations in India, China, Indonesia, and Vietnam; and prawn aquaculture projects in India, Vietnam, Indonesia, China, and Bangladesh. But it was China that was to provide the greatest opportunities for expansion. The CP Group was the first foreign company to invest in China, following the liberalization of the Chinese economy during the late 1970s. In 1979 it was awarded foreign investment ticket number 0001 for both Shantou and Shenzen special economic zones (Ngui, 2001). The group's first investments in China were made in the sector where it had the greatest experience, i.e., in the establishment of feed mills and integrated poultry operations. By 1994, the group operated seventy-five feed mills and extensive poultry-breeding facilities which turned out 260 million day-old chicks per year. The CP Group also repeated its experience of full vertical integration when it came to operate the KFC franchise in thirteen Chinese cities, involving itself in the whole supply chain, from the production of day-old chicks to the annual delivery of 75.5 million birds to the group's KFC outlets. In addition, the CP Group's poultry operations not only supplied the domestic Chinese market but were also utilized as a source of supply to the Japanese market. From this perspective, the CP Group was able to exploit its regional and global span to maintain a foothold in the lucrative Japanese market at a time when exports from Thailand were coming under pressure from cheaper competitors. In China's industrial sector, the group made major investments in the production of motorcycles, in property, telecommunications, petrol exploration and service stations, and retailing (Burch, 1996; Burch and Goss, 1999; Goss, 2002).

By 1995, the CP Group had emerged as the largest agroindustrial company in Asia, with a reported turnover of US$4.05 billion in 1995 and a workforce of 100,000 (excluding many tens of thousands of contract farmers) in over 250 companies in twenty countries. In order to exploit and coordinate its diverse sources of production and its export markets, the CP Group also established its own trading company, CP Trading, which was based on the *sogo shosha,* or Japanese general trading companies such as Marubeni or Mitsubishi. The CP Group was the largest single foreign investor in China, with an involvement in some 130 joint ventures and investments of US$2 billion (Burch, 1996; Burch and Goss, 1999; Goss, Burch, and Rickson, 2000; Loh, 1994; Post Public Publishing, 1994).

The Asian economic crisis of 1997 had a profound impact on the operations of the CP Group, and over the following year the full extent of the problems confronting the group became obvious. By March 1998, TelecomAsia, the group's main telecommunications company, had debts of US$600 million, while the group's Indonesian operations reported a loss of US$54 million and had to request a one-year deferral on debts of US$400 million. The Hong Kong–based holding company that controlled many of the group's operations, CP Pokphand, also recorded a loss of US$107 million for 1997, and in April 1998 would have defaulted on debts of US$93 million had holders of floating rate notes exercised their option of an early redeeming of their holdings. In May 1998, the CP Group entered into negotiations with its creditors over a debt-restructuring plan, which resulted in a reorganization of the group's assets and operations. The group was forced to sell some of its assets and received US$200 million from Tesco, Britain's largest supermarket chain, for a 70 percent holding in the Lotus hypermarket chain in Thailand; the group also sold its 30 percent stake in PetroAsia to its joint project partners; and it disposed of an oil extraction plant and a wheat flour plant in China (Burch and Goss, 1999, Goss, Burch, and Rickson, 2000; Goss, 2002).

The major response to the crisis was the group's decision to concentrate again on its core agrofood operations and on the production and marketing of value-added processed foods. In 1999, the group merged all of its livestock and agricultural businesses in Thailand and a limited number of its overseas operations into one company called Charoen Pokphand Foods (CPF). Its main activities were located in animal feed and livestock products (poultry, pigs, ducks, cattle, and shrimp). The new agrofood grouping had assets of US$1.153 billion in 1999 and sales of US$1.456 billion; whereas the companies in the new grouping registered a combined net loss of US$44.65 million in 1997, this had been turned into a net profit of US$90.4 million in 2000.

Subsequently, the CP Group began to reassert its position as Thailand's leading multinational corporation, with turnover of US$12 billion in 2001. It began once more to expand its operations in both agrofood and nonfood areas. Following increased demand for poultry products from fast-food and other companies in Thailand and overseas, the group invested US$92 million in a new chicken-processing plant and another US$92 in upgrading existing chicken-processing facilities at a number of sites. By 2001, the group was operating about 100 feed mills in China, up from the seventy-five it had in 1994. The group also undertook a number of initiatives designed to exploit new opportunities in the fast-food sector. In 2001, it sold its 51 percent holding in its KFC franchise operation to KFC International, its joint venture partner, in order to concentrate resources on the chain of chicken restaurants known as Chester's Grill. This operation was established in the early 1990s with the aim of building an international chain of restaurants based on a brand image of healthy food and a range of chicken products not normally available in traditional fast-food outlets. By 2001, there were seventy-eight Chester's Grill outlets in Thailand, and the group planned to expand into China, Malaysia, and other Asian markets (Rungfapaisarn, 2001).

In the retail sector in Thailand, the group reduced its investments in supermarkets and chose instead to concentrate on the expansion of its 7-Eleven convenience stores. It opened 300 new 7-Eleven stores in 2002, making a total of over 2,000, with total sales of 24 billion baht (US$550 million). In its retail operations in China, the group adopted the reverse strategy, undertaking major investments in large supermarkets and shopping malls. By 2002, the CP Group operated nine Lotus supermarkets in China and had plans to expand this number to 100 by 2005. The group also established Shanghai's largest shopping mall, the Super Brand Mall, in 2001. In the nonfood areas, the group entered the mobile telephone market in 2001 when it established TA

Orange, a joint venture between TelecomAsia and the French telephone group Orange (Review Publishing Company, 2001; Nation Publishing Group, 2002 a,b,c).

In summary, the CP Group has experienced very rapid and diversified growth throughout its recent history, and for most of that time it experienced consistent growth at the national, regional, and global levels. However, it was severely affected by the 1997 crisis and was forced to sell a number of major assets and once again concentrate on its core business in the production, processing, and marketing of agrofood commodities. Under this strategy, the group is once again emerging as a major regional and global player, across a variety of sectors. How does this compare with the experiences of the Salim Group of Indonesia?

THE SALIM GROUP

The Salim Group is Indonesia's largest conglomerate, the main component of which is PT Indofoods. The Salim Group is owned by the Liem family; was founded by Liem Sioe Liong, who was born the son of a farmer in Fujian province in China in 1917. He left China in 1937 to join his brothers in Indonesia and at first made a modest living selling cloves, peanuts, soap, and bicycle parts in central Java. In the 1950s, he met a young army lieutenant colonel named Suharto, who was the chief supply officer of the Diponegoro division of the Indonesian Army. Liem began supplying oil and other staples to the military base that came under Suharto's command. He also adopted an Indonesian name, Salim (Froman and Canizzo, 1992; Hiscock, 2000; Mulholland and Thomas, 1999).

The Salim family began to prosper after Suharto became president in 1966. In 1969, Bulog, Indonesia's monopoly food agency which was directly controlled by Suharto and his relatives, granted a monopoly on the import, milling, and distribution of wheat to the Bogasari Flour Mill (BFM), a Salim Group subsidiary. Under this arrangement, Bulog imported wheat and sold it to BFM at a subsidized price. The flour was then milled and most of it sold back to Bulog, with BFM adding on a fee for their service. In this way, BFM became the largest wheat flour producer in Indonesia, the world's largest wheat-buying company, and the world's largest producer and exporter of instant noodles (with an output of 7.5 billion packets annually). The Salim Group also expanded into food processing, timber production and forestry products, milk and dairy products (Indomilk), edible oils, and sugar production and refining. All of these activities were rationalized with the formation of PT Indofoods in 1990, which emerged as Indonesia's largest agrofood company and was listed on the Jakarta Stock Exchange in January 1994 (Business World, 1999; England, 1999; Landler, 1999a,b; Shari, 1999).

As with the CP Group, the Salim Group also came to be involved in the food-retailing sector by investing in supermarkets in Indonesia, Singapore, and New York, through the First Pacific Group, a Hong Kong–based investment company in which the Salim Group had a 65 percent holding. The Salim Group also independently established another supermarket chain, Super Indo, which had fifteen outlets in 1999 and operated the third-largest hypermarket chain in Indonesia, with six outlets. More important than any of these retail operations, though, was the establishment in the early 1990s of a network of convenience stores called Indomaret, a local equivalent of the 7-Eleven type of operation which operated on a franchise system. By 2000, 463 Indomaret outlets were in operation, serving as outlets for many of the food and other products manufactured by companies in the Salim Group. The group also moved into in the fast-food sector by acquiring an interest in the KFC franchises in Singapore, Malaysia, and Indonesia, and in 1990 entered into a joint venture agreement with Singapore Food Industries, a government food wholesaler, to supply poultry products to the group's KFC outlets. By late 1997, there were some 150 KFC outlets in thirty Indonesian cities, sixty-nine of which were located in Jakarta (Singapore Press Holdings, 1991a,b; Dow Jones, 1994; ICN, 1997, 2001b; PT Bina Media Tenggara Co., 1997).

In terms of its nonfood operations within Indonesia, the Salim Group invested in a number of ventures including cement, steel, cars, motorcycles, and trucks, chemicals, television stations, telecommunications, soap manufacture, supermarkets, textiles, real estate, sugar and timber plantations, a shipyard, and two industrial parks. One of the most important initiatives undertaken by the group was its move into banking in the late 1950s and the transformation of the Bank Central Asia (BCA) into the largest and most successful private bank in Indonesia (Froman and Canizzo, 1992).

In the 1980s, the Salim Group also began to expand its operations overseas, through a number of investments in both the food and nonfood sectors. In the Philippines, the group established a joint venture with RFM, the Philippines' second-largest food and beverage company, to form RFM Indofoods, a 60-40 partnership engaged in the production of noodles. Through its involvement in a number of investment companies, the Salim Group also came to acquire an indirect interest in numerous agrofood companies in the Asia-Pacific region. However, most overseas investment by the Salim Group was channelled through the First Pacific Group, a Hong Kong–listed investment company established in 1982. Among the earliest investments made by First Pacific was the acquisition of Hagemeyer, an underperforming Dutch trading company which had its origins in colonial Indonesia. First Pacific transformed it into a major trading company selling branded consumer goods, and by 1997, Hagemeyer was operating in twenty-one countries (mostly in North America and Europe) and had a turnover of US$5.7 billion and net profits of US$184 million. The group also had interests in other marketing and distribution agencies including Berli-Jucker, Malaysia Pharmacy, and Thai Pacific Foods. First Pacific also operated banks in Hong Kong and California and undertook major investments in land and property development in Singapore, Malaysia, and Vietnam. As with the CP Group and other Asian conglomerate companies, China also emerged as a major focus for investment and marketing as the Chinese economy was liberalized over the 1980s and beyond (McGregor, 1992; Clifford, 1993; The, 1995; Labita, 1995; Sim, Singapore Press Holdings, 1995; ICN, 1997; Hastings, 1998).

At the time of the Asian economic crisis of 1997, the Salim Group comprised about 500 companies, generated some US$20 billion in annual sales, and employed 200,000 workers. The group operated in virtually every sector of the Indonesian economy and reportedly accounted for some 5 percent of Indonesia's GDP (Froman and Canizzo, 1992). With the onset of the 1997 crisis, all elements of the group's operations faced a critical situation, but the extent of the crisis varied across sectors. The most serious problem for the Salim Group was the Bank Central Asia, which lost 42 percent of its deposits in seven days. In May 1998, the government took over the BCA in order to increase the bank's liquidity, but the run on the bank, along with an 80 percent devaluation of the rupiah and the obligations of other companies within the group, left the Salim Group with some $5 billion (52 trillion rupiah) of debt. Initially, the group tried to meet these debts through asset sales and put Indofoods and Indocement on the market. But Indofoods had lost 17 percent in value in the collapse of August 1997, and in the uncertain conditions surrounding the economic crisis, there were few buyers for the group's assets (Williams, 1997; Landler, 1999a; Flores, 2000).

As with other countries affected by the Asian economic crisis, the World Bank played a significant role, although it was more extensively involved in Indonesia than any other country. The bank made assistance to Indonesia conditional upon the establishment of an agency that would assume control of the assets of indebted private sector companies in return for injections of capital, the pledging of assets as collateral, or the provision of credit. The Indonesian Bank Restructuring Agency (IBRA) was set up in January 1998, and in September 1998, the government of Indonesia reached an agreement with the Salim Group to assume control of the assets of 107 companies in order to repay the group's debts. These assets were lodged with a holding company, Holdiko, which was legally owned by the Salim Group but whose shares were pledged to IBRA. Elements of the Holdiko could be sold by IBRA, but the Salim Group would continue to manage the companies and could regain control of them if the debts were repaid (Dow Jones, 2000; ICN, 2001a).

However, what was important to the Salim Group was not only its ability to reassert control over key assets, but also its capacity to shift these assets overseas in order to put them beyond the reach of the Indonesian government and to protect them from the popular resentment being expressed against the Salim Group as a result of its close association with the corrupt Suharto regime. In this regard, the Salim Group's Hong Kong subsidiary, First Pacific, was particularly important. Although First Pacific was also pledged to IBRA, its overseas location put it beyond the direct control of the agency. Consequently, First Pacific was able to divest itself of some of its assets, such as Tech Pacific, an Australian computer-software company, Pacific Link, a Hong Kong cell phone operator, and a 40 percent stake in Hagemeyer. These sales raised US$2.7 billion, but since the Salim Group retained operational control of First Pacific, this money was not used to retire debt, but rather to purchase a number of core assets from Holdiko, effectively moving these assets from Indonesia to

Hong Kong. The most important of these was PT Indofoods. In 1999, First Pacific purchased a 40 percent stake in Indofood for $650 million and increased its stake by an additional 8 percent in 2000. Thus the Salim Group (in Indonesia) was left with a stake of only 20 percent in Indofoods, while First Pacific (in Hong Kong) came to hold 48 percent (BusinessWorld, 1999; Soloman, 2000; Extel, 2000).

By late 1999, the financial pressures on the group had started to ease. In fact, despite its short-term problems, the flagship company Indofoods even benefited from the economic crisis because falling incomes among poor people led to increased consumption of noodles as a substitute for rice. The company, which was some US$1 billion in debt at the beginning of the economic crisis, returned to profit in 1999. In the first quarter of 1999 it recorded a net profit of US$17 million, compared with a loss of over US$100 million in same period in 1998, and was proposing to repay remaining debt from existing cash flows. By November 2002 the Salim Group finally settled its debts to IBRA and reemerged as a major force in the Indonesian economy as well as the wider region. Even though Indofoods lost its monopoly over flour imports in 1998 and faced increased competition in sales of instant noodles, the company's share of the market (which is currently growing at 4 percent per annum) only fell from 95 percent in 1998 to 88 percent in 2003 (Khagda, 2002; Dhume, 2002; Granitsas, Cohen, and Dhume, 2003).

SOUTHEAST ASIAN AGROFOOD MULTINATIONALS: A COMPARATIVE ANALYSIS

These histories of two of Southeast Asia's most important multinational agribusiness companies point to four significant variables that may account for their growth and development, as well as their differing capacities to cope with crisis. These variables are (1) the nature of the conglomerate company and its future; (2) the structure of vertical integration which enabled agrofood companies to coordinate production and marketing throughout the whole supply chain; (3) the role of the state and the political system in underwriting the emergence of large multinational agribusiness companies; and (4) the role of an emergent China in providing new sites for production and consumption for agrofood multinationals from Southeast Asian countries. Each of these will be considered in turn.

The Asian Conglomerate Company

As noted earlier, the conglomerate model of organization has been criticized in Western business and academic circles as lacking in focus, involving excessively close relationships to government which smack of "crony capitalism," and requiring the exercise of a wide span of control by managers who, because they are selected on the basis of family connections and networks, lack expertise (Hamlin, 2000). In an interview, Michael Porter (cited in Hiscock, 2000, p. 11) has argued that Asian conglomerates organized around family connections do not operate effectively in a globalizing economy, in which success is determined by adding value rather than diversity. Criticisms of the so-called Asian conglomerate were particularly evident during the 1997 crisis, with many observers arguing that this event would result in the demise of this company form.

In our view, such misplaced predictions occurred because of the many misconceptions surrounding the nature of conglomerate companies. For example, as the Boston Consulting Group (BCG, 2002) notes, large-scale conglomerate companies are not an exclusively Asian phenomenon but exist in a variety of forms in most industrialized countries. Examples of non-Asian conglomerates would include companies such as Marconi, Asea Brown Boveri, BHP Billiton, General Electric, United Technologies, Alstom, Alacatel, and Tyco. The BCG report also found that the conglomerate form is not, by definition, inferior to other organizational models and in some respects can be seen to be superior. For example, when compared on the basis of total shareholder returns, conglomerates in the BCG sample of the world's 300 largest corporations performed better than many of the more focused companies. The report argues that all companies have to cope with diversity in one form or another, and that successful conglomerates can capitalize on their diversity and complexity in a variety of ways. In fact, the report argues that at least one form of diversity—either in the range of businesses operated, in the extent of international engagement in new markets, or in the establishment of additional linkages in the supply chain—is important for the growth and survival of all companies. To this extent, the conglomerate model offers important lessons that the more focused companies would do well to address in the future.

Of course, other factors might distinguish the Asian conglomerate from its Western counterpart, but it is by no means clear that these are the same issues which are the subject of criticisms by Western economists and others. For example, while there is no doubt that the "overseas Chinese" dominate the business sector in many Southeast Asian countries, this varies from country to country and is being challenged from a variety of sources. For example, in recent years, this ethnic base of the business community has broadened to include a large element of "overseas Indian" entrepreneurs (Haley and Haley, 1999). In addition, policies of "positive discrimination" in a number of countries have succeeded in establishing a local capitalist class who owe little or nothing to overseas Chinese entrepreneurs. The San Miguel Corporation in the Philippines is a case in point. Finally, the critique of the family-based company which draws its managers from sons and daughters of the founding members is based on a failure to understand the historical contingency of the role of the family in the emergence and maintenance of capitalist enterprises. In the recent past, the family has everywhere played an important role in the staffing of newly established enterprises; equally important, the evidence suggests that over time, family members cease to be active managers and instead become passive shareholders in a company run by professional managers (Bruland and O'Brien, 1998; Agnelli, 2001). Evidence suggests that over time this will also occur in Asian conglomerates; in the case of the CP Group, it is claimed that professional managers already outnumber family executives (Gilley, 2000; Post Public Publishing, 2001).

Asian Agrofood Multinationals and the Retail Sector

One of the distinguishing characteristics of the Asian agrofood conglomerates is the extent to which they have succeeded in fully integrating their operations, both horizontally and vertically. Both of the companies under review—as well as others not considered here, such as the President Group of Taiwan, and the San Miguel Corporation of the Philippines—are not only involved in agrofood and fiber production, but may also own upstream input suppliers, as well as downstream retail outlets such as fast-food restaurants, supermarkets, minimarts, and hypermarkets. They also frequently own the malls and office blocks in which such establishments are located. This capacity to fully integrate both vertically and horizontally has major benefits. For example, as noted earlier, the CP Group's ownership of the chain of 7-Eleven retail outlets not only enables it to market its own food products, but also enables consumers to order a telephone from a CP company, or pay their TelecomAsia telephone bill at a CP-owned 7-Eleven outlet. The capacity to integrate vertically and horizontally is particularly important, because it not only provides opportunities for greater flexibility and profitability, but also allows Asian agrofood multinationals to avoid the control exercised by downstream marketing agencies, such as supermarkets and fast-food outlets, which has proven to be a significant problem for their counterparts in the West (Hughes, 1996; Burch and Goss, 1999).

This issue of manufacturer/retailer relationships, and the distribution of power between these actors in the agrofood chain, is an important topic which has come to be analyzed by a number of researchers in recent years (see e.g., Winson, 1993; Hughes, 1996; Burch and Goss, 1999; Heffernan, 1999; Heffernan and Hendrickson, 2002; Hendricksen et al., 2001). In his earlier studies into vertical integration and concentration in U.S. agriculture, Heffernan (1999) pointed to the acquisitions, mergers, joint ventures, partnerships, contracts, or other strategic alliances that had concentrated production and processing facilities within a series of "food chain clusters" dominated by a few large companies such as Cargill or Conagra. The main focus of this work was the effect of such concentrations on the food system as a whole, and in particular on the farmers who produced raw material inputs to the processing companies. However, in later work (Heffernan and Hendricksen, 2002; Hendricksen et al., 2002) the analysis was broadened to include an evaluation of the effects of the concentration occurring in the retail sector, through the mergers and acquisitions among supermarket chains at the national and global levels. When the retail sector is included in the chain of relationships that make up the agrofood system, it becomes clear that the production and processing companies, which were previously seen to be the most powerful and dynamic element within a commodity chain, were themselves coming to be dominated by the retail sector. The rapid concentration in retailing, combined with the huge purchasing power that large chains exert, has led to a shift in the relationships between producer and retailer, with the balance of power shifting toward the latter:

> Kroger, Albertson's, Wal-Mart, Safeway and Ahold USA account for 42 percent of retail food sales in the United States, whereas in 1997 the top five food retailers had only 24 percent of the market. . . . Trends suggest six or fewer global food retailers will evolve over the next few years. Most of them will likely be European-based transnational firms, such as Tesco (UK), Ahold (Netherlands) and Carrefour (France). . . . Retailers are now in a position to dictate terms to food manufacturers who then force changes back through the system to the farm level. (Hendrickson et al., 2001)

However, while a company such as Cargill has been able to vertically integrate to some degree, starting as an upstream supplier and trader and integrating along the meat and grains chains to the point of producing prepared frozen meals, no processing company has demonstrated an ability to fully vertically integrate along the whole commodity chain and move into the now-dominant retail sector (Wilkinson, 2002).[1] Although no processing company in the industrialized countries has moved into retailing, the supermarket chains and fast-food outlets have been moving back into the production process, for example, through demands that processors supply low-price supermarket brands or "private labels" which compete directly with the branded products of the food manufacturers themselves. Equally important is the fact that the supermarkets decide the terms and conditions by which agrofood companies are able to market their products, by allocating shelf space and determining positioning of products. In short, because of its structural dominance, the retail sector has come to exert wide control over the structures of production and is increasingly important in determining what should be produced, in what quantities, and to what standards and specifications (Hughes, 1996; Burch and Goss, 1999; Wilkinson, 2002).

Clearly, the ability to integrate agrofood production and retailing functions within the same group represents an enormous advantage which Asian agrofood companies possess and Western agrofood companies do not. Asian companies such as the CP and Salim Groups have fully vertically integrated their operations by establishing supermarkets, convenience stores, and fast-food outlets that specialize in the sale of products produced by other companies in the same group. The ability to do this derives in part from the fact that companies such as the CP and Salim Groups were late entrants into the modern agrofood system and became major players at a time when supermarkets and fast-food chains were only just emerging in the less-developed countries. This trend is likely to continue in the future as Asian agrofood companies continue to expand their retail activities on their own account, and as globalizing retail chains such as Tesco, Ahold, and Carrefour move into new markets in Asia and continue their practice of establishing joint ventures with the large, well-established local conglomerate companies which already possess local experience in the retail sector.[2]

The ability of Asian agrofood companies to fully vertically integrate along the whole commodity chain has also meant that the production sector in Thailand and elsewhere has not been squeezed by the domestic retail sector to the extent that it has in the West. Moreover, vertical integration within the agrofood production and consumption sectors in Asia provides an additional benefit, which was demonstrated during the Asian economic crisis of 1997. Even though consumers were spending less as a result of the crisis, food remained an essential item in the domestic budget, and companies with a major interest in food would always experience demand for their products. But more than that, the ownership of food retail outlets meant that companies such as the CP Group were able to rely on the very substantial cash flows that passed through the fast-food outlets and supermarket chains they owned and operated. Cash flow is critically important in any crisis; among other factors, it means that a fully vertically integrated company is in a stronger position when it comes to servicing its debts and reassuring its suppliers (e.g., the processors and farmers who are part of the group and need to be supported). In this regard, the diversified nature of the conglomerate company was a positive factor which helped rather than hindered during the 1997 crisis.

State and Politics in the Emergence of Agrofood Multinationals

Theorists including Dunning (1981) and Bende-Nabende and Slater (n.d) point to the role played by the state in establishing the conditions under which TWMNEs emerge, but another layer of explanation goes beyond issues of effective economic management and policy. There is, for example, the particularistic and preferential behavior of the kind displayed by the Suharto regime toward the Salim Group in the provision of licences and monopolies, which enabled Liem Sioe Liong to create the largest agribusiness company in Indonesia. Without such political leverage, it is doubtful whether the Salim Group could have achieved its preeminent position within the Indonesian economy. There have also been extremely close ties between the CP Group and successive Thai governments, although there is no suggestion of the existence of the corrupt practices that characterized the relationship between the Salim Group and the Suharto government. Nevertheless, Dhanin Chearavanont, other family members, and senior executives of the CP Group have had close ties with most of Thailand's political parties, the bureaucracy, and the military. For example, in 1988, Chearavanont was appointed a senator in the nonelected upper House of Parliament, established by the military junta—the National Peace Keeping Council (NPKC) headed by General Suchinda Kraprayoon—which overthrew the democratically elected government of Chatichai Choonavan in 1991. Following the collapse of the junta after the massacre of protestors in the street of Bangkok in 1992, ex-General Suchinda was given employment within the CP Group. Many other examples of equally close ties have been documented (Nation Publishing Group, 2001, 2002d). The operation of a revolving door, which sees group employees moving into senior government positions and then back to employment within the group, not only confers a high level of access to senior policymakers, but also provides the group with an input into the framing of legislation which impinges directly on the company operations and indirectly on the corporate profitability. Significantly, the group was able to achieve a similar level of access in China, following the appointment of Dhanin Chearavanont as an adviser to the Chinese government. This not only highlights the significance of China to the fortunes of the CP Group, but also reflects the regard in which the CP Group—as the first and largest foreign investor in China—is held among the leadership of that country (Goss and Burch, 2001; Goss, 2002).

The China Dimension

China has come to be a critically important market for the agrofood multinationals emerging out of the Southeast Asia region. In the case of the CP Group, China is important as both a major new market and a low-cost production site from which a wide range of products for sale in Thailand and elsewhere can be sourced. Poultry is the most important commodity in this regard, and the company's chicken-raising operations in China have been used to supply poultry products to Japan in place of the more traditional, and sometimes more expensive, sites in Thailand. One of the CP Group's poultry operations, the Shanghai Dajiang (Group) Limited, turned out 80,000 tonnes of chicken products in 2001, with 25,000 tonnes being exported, mainly to Japan. However, in June 2002, the CP group reduced its holding in Shanghai Dajiang from 43 percent to 25 percent, partly in response to a ban on imports by Japan and South Korea following an outbreak of bird flu, and partly as a consequence of the reappraisal of the group's China strategy, which involved a consolidation of existing assets in order to focus on core businesses in food production, processing, and retailing. Nevertheless, the CP group appears committed to a strategy of maximizing the returns from integrating the production and retail activities of the group in China. By way of contrast, the involvement of the Salim Group in China appears not to be focused on exploiting the possibilities for vertical integration in production and consumption, or using China as a cheap production site by investing in the agrofood or retail sectors. Instead, the Salim Group's engagement with China—which was never as extensive as that of the CP Group—appears to have focused on indirect investments in property developments. For example, the Salim Group was reported to have invested some US$89 million in Fujian province in offices, apartments, and factories, and was part of a nine-member consortium of Indonesian and Singaporean companies that invested US$120 million in the Yuan Hong Industrial Park in 1995 (*China Securities Bulletin,* 2001).

CONCLUSIONS

This chapter has attempted to analyze and evaluate the experience of a small number of multinational agribusiness companies that have emerged from Southeast Asia in recent years, which in many respects challenge a number of the assumptions drawn from the experience of Western agrofood companies. Most significant perhaps, the companies under discussion here have succeeded in vertically integrating along the whole commodity chain in a way that no Western agrofood multinational has been able to do. To this extent, Asian agribusiness companies have been able to avoid the conflicts that have characterized the relationships between food-processing companies and the retail sector in recent years. This does not, of course, guarantee that the farmers who are operating downstream from the retail sector will not be subjected to the same pressures as their Western counterparts, since farmers everywhere are likely to be squeezed by the pressures emanating from a competitive retail sector. Contract farmers in Thailand producing raw material inputs for the CP Group may find themselves in as weak a bargaining position as contract farmers in Australia or England producing peas for a large global processing company (Burch, Rickson, and Annels, 1992; Burch and Rickson, 2001).

The case study material also suggests that although Asian agrofood multinational companies clearly have much in common—such as the extent of vertical integration and their close ties to the state and the political system—there are also significant variations which lead us to reject the overgeneralizations that the use of concepts such as "Asian conglomerates" implies. In terms of overseas investment, for example—which is where this discussion started—the strategy adopted by the CP Group in China was very different from that of the Salim Group. The CP Group has undertaken a large amount of FDI in China, which was clearly related to an overall strategy of regional expansion designed not only to service the China market for value-added agrofood products, but also to use China as a low-cost production base to service markets in Japan and wider global markets. The Salim Group, on the other hand, tended to invest (both directly and indirectly) in the property and service sectors in China, and eventually came to use its major investment vehicle, the Hong Kong–based First Pacific company, as a means of moving ownership of the group's main agrofood company, Indofoods, from Indonesia to the Chinese mainland.

In summary, the attributes of many Asian agrofood companies, as well as the differences which distinguish them from one another, are important not only in terms of what they tell us of the experience of such firms operating in the third world context, but also for the insights they offer in helping us to understand the experience of agrofood companies in the industrialized countries. A program of comparative analysis within countries, between companies, and across national boundaries would yield many insights and much important data and add significantly to our understanding of the forces shaping the emerging global agrofood system.

NOTES

1. In an analysis of growth in food and drink (F&D) multinational companies over the period 1977-1994, Rama (1998) analyzed a sample of sixty-four F&D companies. All of these were involved in the processing sector, but about one-third of the sample was also involved in the retail sector. Although this seems to suggest a potentially higher level of vertical integration than is indicated in this chapter, it is important to understand the nature of this involvement and whether this potential was, in fact, translated into actual integration throughout the commodity chain. As Rama (1998) indicates, much of this vertical integration would have been accounted for by the quite unique situation whereby breweries owned and operated public houses in the United Kingdom and elsewhere. Without access to the primary data used by Rama (1998), it is also possible to suggest that the apparently high level of integrated production and distribution reflected the fact that Britain's cooperative movement was for a long time a major force in the food system. If such is the case, then it is possible to reconcile the data in Rama (1998) with our position which argues that in terms of the agrofood commodity chain in general, the processing/manufacturing sector in Western countries has not moved into the retail sector in order to achieve full vertical integration.

2. In Thailand, Tesco was initially associated with the CP Group, while in Indonesia, Carrefour entered into a partnership with the Tigaraksa Group, another Indonesian conglomerate company engaged in the production of baby food, pharmaceuticals, garments, electronic goods, packaging, and steel, in property and real estate development, and in trading and general distribution. In Taiwan, Carrefour came to be associated with the Uni-President Group. (See Froman and Canizzo, 1992; ICN, 1999).

REFERENCES

Agnelli, G. (2001). Family business and the growth of capitalism. Paper presented at the Twelfth Annual Conference of the Family Business Network, Rome, October 4.

Bende-Nabende, A. and Slater, J.R. (n.d.). *Government Policy, Industrialisation, and the Investment Development Path: The Case of Thailand.* Birmingham Business School Working Paper Series, University of Birmingham.

Boston Consulting Group (2002). *Conglomerates Report 2002.* Sydney: Author.

Bruland, K. and O'Brien, P. (1998). *From Family Firms to Corporate Capitalism.* Oxford: Oxford University Press.

Burch, D. (1996). Globalization agriculture and agri-food restructuring in South East Asia: The Thai experience. In D. Burch, R. Rickson, and G. Lawrence (Eds.), *Globalization and Agri-Food Restructuring: Perspectives from the Australasia Region* (pp. 323-344). Aldershot, UK: Avebury.

Burch, D. and Goss, J. (1999). An end to Fordist food? Economic crisis and the fast food sector in Southeast Asia. In D. Burch, J. Goss, and G. Lawrence (Eds.), *Restructuring Global and Regional Agricultures: Transformations in Australasian Agri-food Economies and Spaces* (pp. 87-110). Aldershot, UK: Ashgate Publishing.

Burch, D. and Rickson, R.E. (2001). Industrialised agriculture: Agribusiness, input-dependency, and vertical integration. In S. Lockie and L. Bourke (Eds.), *Rurality Bites: The Social and Environmental Transformation of Rural Australia* (pp. 165-177). Sydney: Pluto Press.

Burch, D., Rickson, R.E., and Annels, H.R. (1992). Agribusiness in Australia: Rural restructuring, social change, and environmental impacts. In K. Walker (Ed.), *Australian Environmental Policy* (pp. 19-40). Sydney: New South Wales University Press.

BusinessWorld, Inc. (1999). SMC out of the picture in food deal. *Business World, Philippines,* June 23, p. 8.

China Securities Bulletin (2001). June 13.

Clifford, M. (1993). A question of loyalty. *Far Eastern Economic Review,* April 29, p. 54.

Dhume, S. (2002). Review 200-Indonesia-Indofood uses its noodles. *Far Eastern Economic Review,* December 26, p. 68.

Dow Jones (1994). Asia-Pacific brief: Fast food Indonesia. *Asian Wall Street Journal,* November 29, p. 5.

Dow Jones (2000). Indonesia Indofoods seeks to buy Salim noodle plant. *Dow Jones International News,* May 24.

Dunning, J.H. (1981). Explaining the international direct investment position of countries: Toward a dynamic or developmental approach. *Weltwirtschaftliches Archiv,* 119: 30-64.

Dunning, J.H. (1988). *Explaining International Production.* London: Unwin Hyman.

Dunning, J.H. and Hamdani, K.A. (Eds.) (1997). *The New Globalism and Developing Countries.* Tokyo: United Nations University Press.

Dunning, J. H., van Hoesel, R., and Narula, R. (1996). *Explaining the "New" Wave of Outward FDI from Developing Countries: The Case of Taiwan and Korea.* Maastricht Economic Research Institute on Innovation and Technology (MERIT), Research Memoranda 009, Maastricht.

England, V. (1999). Jury out on whether Suharto favourite can survive. *South China Morning Post,* March 5, p. 7.

Extel (2000). First Pacific shareholders approve increase in Indofoods to 48%. *Extel Company News,* October 18.

Flores, S. (2000). SMC interested in two domestic food companies. *Business World,* Philippines, March 14, p. 7.

Froman, L. and Canizzo, M. (1992). *The Salim Group.* Fontainebleau, France: European Institute of Business Administration–Euro-Asia Research Centre (INSEAD-EAC).

Gilley, B. (2000). Showing the way. *Far Eastern Economic Review,* February 10, pp. 45-46.

Goss, J. (2002). Fields of inequality: The waning of national developmentalism and the political economy of agribusiness, Case studies of development and restructuring in Thailand's agri-food sector. Unpublished doctoral thesis, Griffith University, Brisbane, Australia.

Goss, J. and Burch, D (2001). From agricultural modernisation to agri-food globalisation: The waning of national development in Thailand. *Third World Quarterly,* 22(6): 334-351.

Goss, J., Burch, D., and Rickson, R.E. (1998). Global shrimp activism: Trends or tendencies? *Culture and Agriculture,* 20(1): 513-530.

Goss, J., Burch, D., and Rickson, R. (2000). Third world corporate transnationalism: The case of Charoen Pokphand, Thailand and the global shrimp industry. *World Development,* 28(3).

Granitsas, A., Cohen, M., and Dhume, S. (2003). The instant noodle war. *Far Eastern Economic Review,* January 9, p. 42.

Haley, G. and Haley, U. (1999). Weaving opportunities: The influence of overseas Chinese and overseas Indian business networks on Asian business operations. In Richter, F.J. (Ed.), *Business Networks in Asia: Promises, Doubts and Perspectives* (pp. 149-170). Westport, CT: Quorum Books.

Hamlin, M.H. (2000). *The New Asian Corporation: Managing for the Future in Post-Crisis Asia.* San Francisco: Jossey-Bass.

Hastings, K. (1998). Asset sales boost First Pacific's net, raises questions about future. *Dow Jones Online News,* March 31.

Heffernan, W. (1999). *Consolidation in the Food and Agriculture System.* Denver, CO: National Farmers Union.

Heffernan, W. and Hendrickson, M. (2002). Multi-national concentrated food processing and marketing systems and the farm crisis. Paper presented to the Annual Meeting of the American Association for the Advancement of Science, Boston, February 14-19.

Hendrickson, M., Heffernan, W., Howard, P., and Heffernan, J. (2001). *Consolidation in Food Retailing and Dairying.* Denver, CO: National Farmers Union.

Hiscock, G. (2000). *Asia's New Wealth Club.* London: Nicholas Brealey Publishing.

Hughes, A. (1996). Forging new cultures of food: Retailer-manufacturer relations. In N. Wrigley and M. Lowe (Eds.), *Retailing, Consumption and Capital: Toward a New Retail Geography* (pp. 90-115). London: Longman.

ICN (1997). Fast food franchise business: Its conditions and developments. *Indonesian Commercial Newsletter,* December 22, p. 7.

ICN (1999). Tiga Rakso Group: Reinforcing core business through hypermarket. *Indonesian Commercial Newsletter,* April 27, p. 45.

ICN (2001a). Divestment of Holdiko's assets worth RP7.1 trillion. *Indonesian Commercial Newsletter,* July 10, p. 38.

ICN (2001b). Supermarkets blocked by minimarkets and hypermarkets. *Indonesian Commercial Newsletter,* July 10, p. 7.

Kagda, S. (2002). Salim Group settles its 52t rupiah debt with IBRA. *Business Times,* Singapore, November 28.

Khan, K. (Ed.) (1987). *Multinationals of the South.* London: Francis Pinter.

Kumar, K. and Mcleod, M. (Eds.) (1981). *Multinationals from Developing Countries.* Lexington, MA: D.C. Heath and Company.

Labita, A. (1995). US trading for food, beverages firm RFM's depositaries. *Business Times,* Singapore, November 29.

Lall, S. (1983). *Third World Multinationals.* Chichester, UK: John Wiley.

Landler, M. (1999a). An Asia empire still on a seesaw: Gains and setbacks in reviving an Indonesian giant. *The New York Times,* April 28, p. C4.

Landler, M. (1999b). Year of living dangerously for a tycoon in Indonesia. *The New York Times,* May 16, p. 1.

Lecraw, D. (1993). Outward direct investment by Indonesian firms: Motivations and effects. *Journal of International Business Studies,* 24(3): 589-600.

Loh, H.Y. (1994). Charoen Pokphand picks Suchinda to chair subsidiary. *Business Times,* Singapore, November 30.

McGregor, J. (1992). Fujian gets boost from overseas Chinese. *Asian Wall Street Journal,* June 4, p. 1.

Mulholland, J.P. and Thomas, K. (1999). The price of rice. *Inside Indonesia,* 58: 23-24.

Nation Publishing Group (2001). The board, managers and lollipop men overseeing Thaksinland. *The Nation,* Bangkok, February 20.

Nation Publishing Group (2002a). Capital raising: TA plans share sale. *The Nation,* Bangkok, August 9.

Nation Publishing Group (2002b). CP to focus on 7-Eleven. *The Nation,* Bangkok, November 6.

Nation Publishing Group (2002c). Lotus supercentres: CP ready for China expansion. *The Nation,* Bangkok, December 16.

Nation Publishing Group (2002d). Wattana not a CP man. *The Nation,* Bangkok, November 8.

Ngui, C.Y.K. (2001). The giant awakens. *Malaysian Business,* December 1, p. 25.

Peng, M., Au, K., and Wang, D. (2001). Interlocking directorates as corporate governance in third world multinationals: Theory and evidence from Thailand. *Asia-Pacific Journal of Management,* 18: 161-181.

Post Public Publishing Co. (1994). Suchinda position defended. *Bangkok Post,* November 29, p. 21.

Post Public Publishing Co. (2001). Third generation meets the challenge. *Bangkok Post,* October 23.

PT Bina Media Tenggara Co. (1997). Salim Group quits PT Gelael. *Jakarta Post,* May 6, p. 12.

Rama, R. (1998). Growth in food and drink multinationals, 1977-94: An empirical investigation. *Journal of International Food and Agribusiness Marketing,* 10(1): 31-52.

Review Publishing Co. (2001). Thai brand mall. *Far Eastern Economic Review,* December 27, p. 6.

Richter, F. J. (Ed.) (1999). *Business Networks in Asia: Promises, Doubts and Perspectives.* Westport, CT: Quorum Books.

Rungfapaisarn, K. (2001). No Chicken Little in this plan. *The Nation,* Bangkok, August 22.

Shari, M. (1999). Anthony Salim's comeback may be coming apart. *Business Week,* May 3, p. 28.

Sim, W.C. (1995). Salim Group mounts takeover bid for listed food group QAF. *Straits Times,* April 13.

Singapore Press Holdings (1991a). KMP moves into corporate limelight. *Business Times,* Singapore, January 25, p. 11.

Singapore Press Holdings (1991b). KMP's managing director joins Inno-Pacific board. *Business Times,* Singapore, January 9, p. 3.

Singapore Press Holdings (1995). Singapore joint-venture industrial park in Fujian opens. *Straits Times,* May 26.

Soloman, J. (2000). Salim Group: Nearly sunk, now expands across Asia. *Asian Wall Street Journal,* January 13, p. 1.

The, H.L. (1995). Another US$60M for China industrial park. *Business Times,* Singapore, November 4.

Ting, W. (1985). *Business and Technological Dynamics in Newly Industrializing Asia.* Westport, CT: Quorum Books.

Wells, L.T. (1983). *Third World Multinationals: The Rise of Foreign Direct Investment from Developing Countries.* Cambridge, MA: MIT Press.

Wilkinson, J. (2002). GMOs, organics and the contested construction of demand in the agrofood system. *International Journal of the Sociology of Agriculture and Food,* 10(1): 1-24.

Williams, L. (1997). Government arms stems panic selling. *Sydney Morning Herald,* August 18, p. 26.

Winson, A. (1993). *The Intimate Commodity: Food and the Development of the Agro-Industrial Complex in Canada.* Aurora, Ontario: Garamond Press.

World Bank (1994). *The East Asian Miracle: Economic Growth and Public Policy.* New York: Oxford University Press.

Chapter 9

Multinational Firms
in the Brazilian Food Industry

Elizabeth Farina
Cláudia Viegas

INTRODUCTION

Investigating Brazil's importance in attracting investments from multinational enterprises (MNEs) and determining which sectors are leaders in this movement is of fundamental importance in assessing the recent economic development of Brazil and Latin America. Economic stabilization, achieved with the Real Plan,[1] the growing liberalization of the economy, and the expansion of the consumer market have caused many multinationals in the food area to choose Brazil as headquarters for their investments in MERCOSUR (the Southern Cone Common Market). Empirical evidence shows that mergers and acquisitions (M&A) have been the most common strategy among firms interested in entering the Brazilian food market. In this manner, the multinational manages to establish its brand name in the new market, adapting to local consumption habits and gaining market share rapidly. This has altered the national competitive scenario and increased the importance of Brazil in MERCOSUR. This chapter assesses the impact of this entry of foreign capital on the Brazilian food industry as well as its importance, evolution, and trends.

Brazil is a strong player in the global agrofood business sector, with 170 million inhabitants, a size of 8.5 million km^2 and a GDP of US$511 billion in 2001 (Brazilian Institute of Geography and Statistics [IGBE], 2002). According to the Food and Agriculture Organization of the United Nations (FAO/UN), nearly 19 percent of the world's arable land is in Brazil, but the country uses just 10 percent of this area. Moreover, 19 percent of the planet's water is in Brazil. With a wide range of latitudes and reasonably well-distributed rainfall throughout the year, the country is able to produce a wide range of products varying from coffee and soybeans to apples, pears, melons, and grapes. Consequently, agricultural production and productivity have large growth potential through the incorporation of new areas and technology adoption (Farina and Nunes, 2002).

Brazil is the largest producer of fresh oranges and exporter of concentrated frozen orange juice. It ranks second in soybean production and exports following the United States. It is the third biggest corn producer behind the United States and China; the second biggest tobacco producer and exporter behind the United States; second in sugar cane and raw sugar production following India; and the largest exporter of sugar cane, coffee, and alcohol. Brazil is second behind the United States as producer and third as exporter of beef, and second in broiler production and exports.

By the end of the twentieth century, Brazil accounted for 78 percent of the population of MERCOSUR. Its GDP was 62 percent of the MERCOSUR total, although it had only one-half the per capita GDP of Argentina and suffered from poor income distribution. Brazil's exports were 62 percent and its imports 60 percent of the MERCOSUR totals; 43 percent of intrabloc exports and 46 percent of intrabloc imports were Brazilian (Farina, 1999).

As regards its internal market, Brazilian food consumption growth since the stabilization plan in 1994 (Real Plan) has been astonishing when compared to European and U.S. markets. Dairy products grew 25 percent in volume, while yogurt grew more than 80 percent. Moreover, the percentage of average family expenditure on dinners and lunches away from home grew 80 percent, while consumption of prepared foods almost doubled, as did soft drinks and hot and cold beverages.

In sum, Brazil is an important case to be analyzed due to the large consumer market and comparative and competitive advantages in total production. It is not surprising that the country has attracted a considerable amount of foreign direct investment (FDI) to the food sector.

The chapter is divided into five sections. The next section presents a general overview of foreign direct investments in Brazil and Latin

America. The section reviews the literature on entry of foreign capital into this region, mainly in the 1990s. The objective is to show the position occupied by the area in attracting investments from multinationals, highlighting the main receiving countries. Greater emphasis is given to the food industry, showing the countries that most stand out in attracting these investments, and demonstrating the Brazilian leadership in attracting FDI to Latin America. Subsequent sections then study in greater detail mergers and acquisitions, because this is the entry strategy predominantly adopted by the food multinationals entering Brazil (concerning the importance of M&A as an entry mode, see Chapter 1).

Following the same line of argument, a later section examines the Brazilian food market as the main attraction for FDI. Changes in consumption habits, given the economic stability achieved with the Real Plan, and the size of the national market were the main factors that placed the country in the forefront of the strategies of agrofood multinationals. The fifth section investigates the impact of these investment flows on Brazilian industry. The main variables studied relate to the competitive environment, employment, and new competition strategies. That is, the large number of mergers and acquisitions with the presence of foreign capital in the food industry has an impact on the levels of investment (in addition to ownership transference) and employment, which heavily alters the patterns of competition.

Finally, the conclusion presents an analysis of the multinational firms in the Brazilian food industry, showing the important role of this sector in attracting foreign direct investments to Latin America. The consumer market and relative economic and political stability in addition to Brazil's large capacity to produce agriculture products cause multinational firms in the food business to see Brazil as the best option for participation in Latin America. Thus Brazil is chosen as the headquarters for these investments. The entry of foreign capital, especially through M&A, has significantly changed the patterns of competition in the Brazilian food sector, guaranteeing the sector greater dynamism and competitiveness.

GENERAL OVERVIEW OF FOREIGN DIRECT
INVESTMENTS IN BRAZIL AND LATIN AMERICA

According to data from the United Nations Conference on Trade and Development (UNCTAD) (2002), Latin America attracted $56.8 billion in FDI in 2000, of which 78 percent went to MERCOSUR countries (Argentina, Brazil, Paraguay, and Uruguay). The increased importance of the MERCOSUR countries as recipients of FDI should be stressed. In 2001, Brazil was, for the sixth consecutive year, the leader in bringing FDI to the region, attracting US$22.5 billion. This number is rather small if compared with the inflows of FDI to Brazil in the previous years. In 2000, the amount was US$32.779 billion.[2]

The decline of FDI in Brazil in 2001 followed the worldwide decrease in investments in all countries, except African countries, in that year. This global downturn can be explained mainly by the slowdown of world economic growth in 2001. In addition, Brazil reduced the pace of privatization, particularly in the services sector, such as the telecommunications industry.

The movements of capital observed in Latin America suggest that foreign investors give more importance to the stability of the institutional and political structure than to conjunctural macroeconomic crises. Brazil and Chile received the most investments, as compared to Paraguay, Colombia, Ecuador, and Venezuela which all saw a reduction in the flow of FDI. There is an interesting difference between the role and the structural consequences of FDI in Mexico and the Caribbean and those of the other Latin America countries. Recent FDI in South America has been concentrated in nontradable services, manufactured products for the local market, and intensive activities in natural resources. The consequence is that FDI has not had a strong impact on the export structure of these countries, highly concentrated in commodities and basic natural resources. In countries such as Mexico and the Caribbean, multinational firms have used the region as an export platform to the North American market, improving the competitive position of these countries and redirecting their productive and commercial structure toward the foreign market. The multinationals chiefly exported automobiles, electronics, and textiles.

Recent data show that industrialized countries have a growing interest in investing in Latin American countries. Table 9.1 shows the increase of FDI in Latin America, while Table 9.2 indicates the values

TABLE 9.1. Flows of FDI announced in Latin America (millions of U.S. dollars).

Countries	Inflows in 1998	Inflows in 1999	Inflows in 2000	Inflows in 2001
Argentina	6,848	24,134	11,152	3,181
Bolivia	952	985	693	647
Brazil	28,856	28,578	32,779	22,457
Chile	4,638	9,221	3,674	5,508
Paraguay	336	66	96	152
Uruguay	164	239	285	320
Sum	41,794	63,223	48,679	32,265
Total	51,886	70,880	56,837	40,111
Sum/Total as %	80.55	89.20	85.65	80.44

Source: UNCTAD, 1999, 2000, 2001, 2002.

TABLE 9.2. Mergers and acquisitions (millions of U.S. dollars).

Countries	Sales in 1998*	Sales in 1999*	Sales in 2000*	Sales in 2001*
Argentina	10,396	19,407	5,273	5,431
Bolivia	180	232	19	–
Brazil	29,376	9,357	23,013	7,003
Chile	1,595	8,361	2,929	2,830
Paraguay	11	–	65	67
Uruguay	36	–	27	36
Sum	41,594	37,357	31,326	12,537
Total	46,834	39,033	35,584	16,174
Sum/Total as %	88.81	95.71	88.03	77.51

Source: UNCTAD, 1999, 2000, 2001, 2002.
*Registered in the country of the target firm.

of M&A. A strong concentration of FDI within the extended MERCOSUR[3] benefits Brazil, Argentina, and Chile (Table 9.1). Data on bilateral investments between Brazil and Argentina (see Table 9.3) show that Argentinean FDI in Brazil was predominant, especially by the end of the 1990s.

According to J. H. Dunning

> the multinational enterprise is assuming, more and more, the role of orchestra conductor, in relation to various activities of production and transactions, which occur within a "network" of transnational relationships, both inside and outside the companies, and that may or may not include an investment of capital, but whose objective is to promote its global interests. (Chesnais, 1996, p. 68)

From 1990 to 1995, U.S. direct investments in the western hemisphere expanded rapidly (Table 9.4). Canada is the largest host country in the hemisphere, followed by Mexico, Brazil, and Argentina. The ranking is the same for total FDI and for the food industry.

The growth of American investment in Latin American countries can be attributed to several factors, such as the following:

- In recent years, many countries liberalized their rules on foreign trade.
- Population growth has increased the demand for food overall.
- Increase in income has increased the demand for processed foods and also the demand for differentiated foods (dietary, healthier foods).
- Greater stability of many countries has made the environment more favorable to the increase of investments, both national and foreign.
- Regional trade agreements, such as NAFTA (Canada, United States, and Mexico) and MERCOSUR (Argentina, Brazil, Chile, Paraguay, and Uruguay), have encouraged investors (Bolling, Neff, and Handy, 1998).

These investments seem to have had beneficial effects on the economy of the host country by contributing to the basis of food production. The production of processed foods in the host country can be obtained at lower costs when compared to the cost of imported foods. At

TABLE 9.3. Bilateral flows of direct investments.

FDI	1990-1997	%	1998-2000	%	Total	%
Investments of Brazilian firms in Argentina	1207	54.3	1529	22.2	2736	30.0
Investments of Argentinean firms in Brazil	1017	45.7	5355	77.8	6372	70.0
Total (bilateral flow)	2224	100.0	6884	100.0	9108	100.0

Source: CEP, 1998, n. 20, Table 8.1, in Bonelli, 2000, p. 18.

TABLE 9.4. Direct investments of the United States in the Western hemisphere.

Country/region	US$ billion				% change		
	1998	1999	2000	2001	1998/1999	1999/2000	2000/2001
Total FDI	1,000.7	1,173.1	1,293.4	1,381.6	17.2	10.3	6.8
Food industry	35.3	34.2	35.9	35.4	-3.1	5.0	-1.4
Total FDI in Canada	98.2	111.7	128.8	139.0	13.7	15.3	7.9
Food industry	5.0	5.8	5.5	4.6	16.0	-5.2	-16.4
Total FDI in Latin America	128.6	174.0	154.5	160.7	35.3	-11.2	4.0
Food industry	9.9	8.0	7.8	8.6	-19.2	-2.5	10.3
Total FDI in Mexico	26.7	32.9	37.3	52.2	23.2	13.4	39.9
Food industry	4.7	3.7	4.4	4.5	-21.3	18.9	2.3
Total FDI in Central America	56.0	69.3	70.5	80.6	23.8	1.7	14.3
Food industry	16.8	4.2	4.9	4.8	-75.0	16.7	-2.0
Total FDI in South America	72.6	77.7	84.0	83.4	7.0	8.1	-0.7
Food industry	4.7	3.8	4.1	3.8	-19.1	7.9	-7.3
Total FDI in Brazil	37.2	37.4	39.0	36.3	0.5	4.3	-6.9
Food industry	2.6	1.3	1.6	1.4	-50.0	23.1	-12.5
Total FDI in Argentina	12.3	15.6	9.4	14.2	26.8	-39.7	51.1
Food industry	1.0	1.17	0.5	0.8	17.0	-57.3	60.0

Source: U.S. Department of Commerce, Bureau of Economic Analysis data. Available at <http://www.bea.doc.gov>.
Note: The Bureau of Economic Analysis defines U.S. direct investment abroad as ownership by a U.S. investor of at least 10 percent of a foreign business. U.S. direct investment abroad is defined as the ownership or control, directly or indirectly, by one U.S. resident of 10 percent or more of the voting securities of an incorporated foreign business enterprise or the equivalent interest in an unincorporated foreign business enterprise.

the same time, the direct investment in the host country creates jobs, increases the gross domestic investment, and can generate export flows, in turn generating foreign exchange for the country. Canada, Mexico, Brazil, and Argentina are responsible for US$9.9 billion (90 percent) of the total US$11 billion FDI of American companies in the western hemisphere (Bolling, Neff, and Handy, 1998). However, as most of the recent FDI in the food industry has occurred through acquisitions, the beneficial effects depend on modernization and expansionary investments. As will be discussed later, there is evidence of positive effects related to production and productive growth in the industry segment as well as in retailing.

The *eclectic paradigm* of international production postulates that the stock of foreign assets owned and controlled by multinational enterprises is determined by (1) the competitive advantages (O) of those firms vis-à-vis those of uninational firms; (2) the extent and nature of the location (L)-bound endowments and markets offered by countries to firms to create or add further value to these competitive advantages; and (3) the extent to which the market for these advantages is best internalized by the firm itself (I), rather than marketed directly to foreign firms. As stated in the *internalization paradigm*, there are advantages of internalizing the market in order to appropriate the full economic rent created by core assets (Dunning, 1995a, p. 81).

The OLI advantages will vary according to country, nature of activity, and firm-specific characteristics. Moreover, the propensity of corporations to engage in foreign production will be more pronounced the greater their relative competitive advantages and the more they find it profitable to create or add value to these advantages themselves from a foreign location (Dunning, 1995a, p. 81).

"One of the specific aspects and also one of the privileges or 'personal advantages' of multinationals is to constitute, between headquarters and subsidiaries, an internal market" (Chesnais, 1996, p. 82). This gives a peculiar character to the sources of competitiveness and to the economic power of a given company. Chesnais agrees that transaction costs explain multinationals. The higher the transaction costs, the more advantageous it is to invest directly instead of exporting or selling licenses. Thus, in addition to the companies creating branches in several foreign countries, they must also be narrowly linked in order to dominate the international internalization of transaction costs (Chesnais, 1996, p. 84).

In Dunning's conception,

> enterprises prefer to replace or not utilize market mechanisms, and distribute their resources around their own control procedures; not only do they gain from this, but other enterprises run the risk of suffering losses (in particular, those who were their clients or suppliers before vertical integration, or their competitors before horizontal integration). Internalization constitutes, therefore, a powerful motivation for purchases and mergers and a precise instrument in the strategy of oligopolists. . . . The logic of internalization, therefore, is that it provides a means, not only of safeguarding the monopolistic advantage of the companies, but also of creating and more importantly of reinforcing this advantage. (Chesnais, 1996, p. 85)

It is a means for the company to maintain its specific advantage on a world scale. Furthermore, it is through internalization that it becomes possible to transform an intangible good—a given item of knowledge, for example—into a property element.

We conclude that international production depends "on the configuration of the O endowment of the firm and the L advantages of the country, as well the degree to which firms perceive that, instead of the market, they have net I advantages by organizing these O and L (potential) endowments within the hierarchy" (Negri, 1997, p. 9). Box 9.1 sums up the general aspects of the three advantages addressed here, the OLI advantages.

According to the OLI analysis proposed by Dunning, the predominance of FDI instead of commerce in the food industry would be justified by the conjunction of the following factors:

1. *O advantages:* The MNE possesses a stock of products that can be easily launched in the host country as already having been developed in the country of origin. In other words, the adaptation costs are smaller than the R&D costs of the foodstuff. Moreover, the technological dominance and the performance on a worldwide scale confer advantages on the cost of product development and of publicity, e.g., the case of McDonalds. The anticipation of market trends that will not yet have happened in the developing countries but already exist in the countries of origin of the MNEs—for instance, the increasing concern with

health (e.g., light and diet food segments)—increase the advantages of the MNEs relative to the domestic firms, as the former already have some know-how on how to react to these changes in consumer preferences.

2. *L advantages:* The countries are well endowed with agricultural raw materials. Therefore, direct production reduces the costs of access to and use of the raw material. The high costs of transportation, in problems with port structures and domestic transportation, make it difficult for the foreign firm to trade, not to mention the perishable nature of food. Nontariff barriers can discourage commerce and stimulate, instead, direct production in the domestic market.[4]

3. *I advantages:* To control long-distance partners implies high costs, especially in countries with high costs of monitoring and enforcing property rights. Moreover, producing in situ gives the MNE the advantage of better knowing local markets and decoding consumer idiosyncrasies, which allows it to launch and adapt its foodstuffs before other international competitors.

Low-income countries that enjoy a greater marginal propensity to consume face significant increases in demand for food in periods of economic acceleration. This was precisely the Brazilian case soon after the success of the 1994 stabilization plan.[5] In addition to the response of demand to the GNP growth that followed the first post–Real Plan years, the redistributive effect of monetary stabilization provoked an unprecedented increase in the volume of foods sold, especially those of greater added value, such as dairy products. This behavior of food demand in these countries functioned as a lure to multinationals, further reinforced if we take into consideration the fact that, as a general rule, the countries of origin of these investments display slower growth in the consumer demand for food. These aspects will be explored in greater detail later, in a specific study of the Brazilian consumer market. Besides being a large agricultural and livestock producer, Brazil has a huge internal market, where income inequalities give room for a wide range of competition strategies, from bulk commodities to high added value products. Therefore, OLI factors are present in the Brazilian case.

BOX 9.1. Strategies Linked to the Behavior of Multinational Firms (Dunning).

Specific advantages of the firm (O)	Advantages deriving from internalization (I)	Variables that affect the options of location (L)
Personal advantages • Ownership of technology and specific endowments (personnel, capital, organization) *Advantages linked to the organization as a group* • Economies of scale; market power as seller and as buyer; access to the markets (of factors and of products) • Prior multinationalization; knowledge of the world market; knowledge of international management; capacity to explore the differences between countries; knowledge of risk management	• Minimization of transaction costs in the acquisition of inputs (including technology) • Reduction of uncertainty • Greater protection of technology • Access to the synergies of interdependent activities • Control of validity and initiatives • Possibility of avoiding or of exploiting government measures (especially fiscal) • Possibility of practicing manipulation of transference prices, predatory price setting, etc.	• Specific resources of the country • Quality and prices of inputs • Quality of the infrastructures and externalities (R&D, etc.) • Costs of transport and communication • Psychological distance (language, culture, etc.) • Trade policy (tariff and nontariff barriers, curtailment) • Protectionist threats • Industrial, technological, and social policy • Subventions and incentives to attract companies

Source: Dunning, J. H., 1988, *Explaining international production,* London: Unwin Hyman, in Chesnais, 1996, p. 86.

M&A: MAIN STRATEGY OF ENTRY INTO THE BRAZILIAN MARKET

The main form of entry of foreign direct investments into Latin America occurs via mergers and acquisitions. Focusing on the Brazilian experience, this section analyzes data on this strategic form (for general aspects on M&A in the multinational agrofood sector, see Chapter 1).

Table A9.1 shows the main M&As in the Brazilian food and beverage industry, from 1988 to August 2001, sorted by acquired and purchasing firm. It is interesting to note that these M&As occur among leading companies, which—as will be seen as follows—adds to the concentration of the sector.

Since 1992, the food and beverage industry has stood out among the sectors with the greatest number of M&As (see Table 9.5). Their

TABLE 9.5. Mergers and acquisitions in Brazil, by industrial sector, 1992 to 1998.

Sector	1992	1993	1994	1995	1996	1997	1998	Total sector*	Share in total, 1992-1998
Food, beverages, and tobacco	12	28	21	24	38	49	36	208	12.6
Financial institutions	4	8	15	20	31	36	28	142	8.6
Chemical and petrochemical	4	18	14	13	18	22	25	114	6.9
Metallurgy and steel	11	13	11	9	17	18	23	102	6.2
Insurance	1	1	8	9	16	24	15	74	4.5
Electrical and electronics	2	7	5	14	15	19	9	71	4.3
Telecommunications	1	7	5	8	5	14	31	71	4.3
Others	23	68	96	115	188	190	184	864	52.5
Total mergers	58	150	175	212	328	372	351	1646	100.0

Source: KPMG, 1998, in Faveret, 1999, p. 3.
*Sum from 1992 to 1998.

increase promoted concentration and denationalization of capital in the food, beverage, and tobacco sector of Brazil. By 2000, eight of the ten largest food companies in the country were multinationals, while in 1994 only five were multinationals (see Table 9.9). Foreign companies made 80 percent of the acquisitions of Brazilian food companies in 1996-1997, with enterprises from Argentina, the United States, and Italy as leading investors. In 1998-2000, firms from nations such as Ireland, Mexico, and Chile joined them.

Table 9.6 shows the evolution of the share of foreign capital in total M&A in Brazil for all industries, from 1994 to the first half of 2000. According to the KPMG,[6] Brazil stands out for the size, diversity, and potential of its economy, turning the market into the most attractive for purchases and entrepreneurial alliances in Latin America. In the first half of 2000, 172 business deals were settled, a number 21 percent higher than those registered in the same period in 1999. In the first half of 2000, the information technology sector led the ranking with thirty-two operations, followed by the petroleum industry with twenty-five M&As, and by the food, beverage, and tobacco industry which saw a total of seventeen deals.

A total of 2,127 M&A operations were realized from 1994 to 2000. Foreign capital was present in 60 percent of these operations. In the year 2000, 34 percent of the foreign investments that entered Brazil through acquisition of national firms came from the United States.

From 1994 to 2000, the food sector had a total of 269 M&As, which accounts for 12 percent of the total for all sectors. Of this vol-

TABLE 9.6. Number of mergers and acquisitions realized in Brazil.

Year	Domestic capital	%	International capital	%	Total
1994	81	46.3	94	53.7	175
1995	82	38.7	130	61.3	212
1996	161	49.1	167	50.9	328
1997	168	45.2	204	54.8	372
1998	130	37.0	221	63.0	351
1999	101	32.7	208	67.3	309
2000*	52	30.2	120	69.8	172

Source: KPMG, in *Gazeta Mercantil Latino-Americana,* August 7-13, 2000, p.8.
*First six months.

ume, 57 percent of the operations involved foreign capital, when foreign companies purchased control of local companies or set up joint ventures to create a third company.

Only in 2000 did the food sector lose the leadership in the ranking of the number of M&As to the information and Internet technology sectors. Business deals in the food sector totaled thirty-six operations, or 10 percent of the total. As proposed by Dunning's framework, the main motives for this restructuring of the sector were the spread of businesses, the consumer market, and the opportunities of increased scale. Moreover, most of the national firms were family owned and presented problems of succession within a scenario of debt and increasing competitive pressure (Farina, 1999, pp. 319-320).

THE BRAZILIAN FOOD MARKET
AS THE MAIN ATTRACTION FOR FDI

The national market deserves emphasis as a factor attracting FDI in the Brazilian food and beverage sector. As incomes grew following the stabilization plan, people increased their consumption of greater value-added products, including such diverse products as cookies and hot dogs (see Table 9.7). The expansion of the consumer markets is, as stated, the leading determinant for the entry of foreign firms in the food sector.

Table 9.8 compares the per capita consumption of meats in Brazil and the United States, showing the growth potential in the former. Given that, in general, the consumer markets of the home countries of MNEs stagnate or show modest growth rates, possibilities for rapid increase of their industrial market share has encouraged the rapid entry of such firms in the food and beverage sector in Brazil.

Given this information, the substantial role of the Brazilian consumer market as a lure for investments from agrofood multinationals is undeniable. However, the issues connected to income distribution and to the economic deceleration of Brazil in recent years must also be taken into consideration for a better characterization of the present configuration of the Brazilian consumer market. We will turn to this next.

TABLE 9.7. Evolution of Brazilian consumption, kilograms per inhabitant per year.

Product	1994	1995	1996	1997	1998	1999	2000	Total growth 1994-2000 (%)	Average annual growth 1995-2000 (%)
Hot dogs	0.53	0.74	0.79	0.94	0.98	1.0	0.97	83.02	9.40
Milk*	11.02	13.41	13.59	13.84	13.05	13.96	13.17	19.51	2.05
Powdered milk	0.65	0.76	0.74	0.78	0.75	0.71	0.71	9.23	0.51
Macaroni	2.97	3.34	3.36	3.36	3.54	3.48	3.31	11.45	1.64
Cookies	2.75	3.43	3.83	4.34	4.75	4.77	4.37	58.91	8.08
Powdered coffee	1.56	1.81	2.01	2.09	2.07	2.09	1.93	23.72	6.41
Chocolate	0.47	0.62	0.74	0.82	0.73	0.66	0.65	38.30	3.87
Table wine*	0.18	0.24	0.26	0.28	0.27	0.36	0.37	105.56	10.75
Soft drinks*	34.4	46.6	48.3	50.6	54.3	57.8	59.7	73.55	7.86
Beer*	28.5	31.8	31.3	31.9	33.3	33.3	32.6	14.39	1.99

Source: A.C. Nielsen.
*Liters per inhabitant per year.

TABLE 9.8. Per capita consumption of meats, 1993 and 1998.

Meat product	Brazil			United States		
	Per capita (kg)			Per capita (kg)		
	1993	1998	Total growth (%)	1993	1998	Total growth (%)
Chicken	18.0	24.6	36.7	21.9	23.04	4.9
Beef	35.6	37.4	5.1	27.9	29.4	5.2
Pork	8.1	9.5	17.3	22.2	22.3	0.6
Total	61.7	71.5	16.0	72.0	74.7	3.7

Source: A.C. Nielsen and ERS/USDA, <www.ers.usda.gov>, April 1, 2002.

The Food Market in Brazil

The Brazilian consumer pattern has been changing through the past few years, a reflection of a range of social and economic changes that took place in Brazil over this period. According to data from IBGE, the total resident population jumped from 119 million people in 1980 to more than 169 million in 2000. In 1990, the life expectancy at birth was 65.75 years; in 2001 this number had risen to 68.82. During the same period, the birth rate fell from 23.5 per 1,000 inhabitants to 19.89. These figures explain the changes in the age composition in Brazil. In 1980, people up to fourteen years of age represented 38.2 percent of the Brazilian population; in 2000, this percentage fell to 29.6 percent of the population. On the other hand, people between fifteen and sixty-four years, who accounted for 57.68 percent of the total population in 1980, amounted to 64.55 percent in 2000. For the population sixty-five years old or over, the percentages rose from 4.01 percent to 5.85 percent between 1980 and 2000. On the other hand, the urban population jumped from 67.59 percent of total in 1980 to 81.25 percent in 2000. These figures contribute to explaining changes in Brazilian consumption patterns, which affected the performance and strategies of the food industry. Moreover, between 1992 and 1999, living conditions improved, though major inequalities still remain (IBGE, 2002).

For instance, national income remains highly concentrated. Brazil ended the 1990s with practically the same income distribution as in 1992. The poorest 50 percent of the population have 14 percent of the country's revenue, whereas the richest 1 percent of the population have 13 percent of the total income. The Gini index oscillated between 0.571 in 1992 and 0.567 in 1999; in other words, there was no significant change.

Yet living conditions improved. Life expectancy rose 2.1 years, the number of homes with basic sanitation increased by 18.1 percent, the monthly average income grew 29.8 percent, and infant mortality fell 22.1 percent. It is important to point out that these figures show variations from one state to another, reflecting the strong regional inequalities of the country. For example, the literacy rate in São Paulo, Rio Grande do Sul, and Santa Catarina is over 80 percent, whereas the national average is just 71.6 percent.

The consequence of inequality for the food industry is negative in the short run, as poverty means low demand. In the long run, however, inequality means that there is still room for rapid income and demand growth.

Low income elasticity of foodstuffs gives this industry the peculiar characteristic of being less sensitive than other economic sectors to economic fluctuations. By contrast, the income distribution elasticity is higher than in other economic sectors. The beginning of the Real Plan, by stabilizing the economy, increased the purchasing power of the lower-income classes. This brought benefits to the food sector, both to basic products as well as to foodstuffs of greater added value. The latter have greater income elasticity than other products and are, therefore, more susceptible to economic cycles. Given the high degree of income concentration in Brazil, firms could not manage to grow extensively, that is, through new customers. Winning consumers is, thus, heavily tied to the capacity to occupy someone else's space.

However, the competition pattern has changed over this period, and new brands have threatened the leaders, including low-price brands. Economic stability protects wages and salaries from inflationary losses, increasing the frequency of shopping trips.[7] This stimulates the consumption of new products, because more frequent visits to the grocery store make it possible to consume smaller quantities in order to try new brands. This, in turn, increases levels of competition. As Brazilian family companies were facing internal tensions over succession, higher competition levels facilitated their acquisition by multinationals.

According to a recent study by A.C. Nielsen, from 1998 to 2000 the leading brands, out of 157 categories of products surveyed, lost market share in Brazil. The most notable losses occurred in nonalcoholic beverages, sweets, and household cleaning products. Only 12 percent of the leaders managed to gain market share in this period, whereas 63 percent lost it. Without doubt, the food sector stands out in this process. Breakfast cereals provide a remarkable example: in 1995, Kellogg's held 72 percent of the market; today the brand has only 47 percent. Nestlé emerged as a major rival in 1995 and has 21 percent of sales at present. In this vein, probably the most striking phenomenon is that out of every 100 packages of cereal consumed in the country, thirty-two are little-known brand names such as

Xereta. Moreover, the number of secondary brand names in powdered chocolate has more than tripled since 1990. In 1995, the Nescau brand held 63 percent of sales; today it holds 52 percent. Even though the leading brand names have lost some of their market share, Table 9.9 shows that, among the ten biggest food enterprises, the numbers of MNEs have increased since 1994.

These data reflect changes in the behavior of the consumer. Once the euphoria of the Real Plan faded, the domestic market no longer demonstrated significant growth. Furthermore, as discussed previously, data from IBGE shows that income distribution in Brazil has not improved significantly in recent years and remains highly concentrated. In periods of economic deceleration we should expect a drop in consumption of products of higher added value, which are much more sensitive to income levels as they display high income elasticity. Thus consumer attention turns to products of lower added value. This shift could give space to cheaper brands which, because they invest less in advertising, enjoy lower costs. There is every indication that this has proven to be a fundamental strategy for the consumer who, being unwilling to pay more just for brand names, instead chooses low prices.

IMPACTS ON THE FOOD INDUSTRY OF BRAZIL

Effects on the Competitive Environment

The pattern of concentration of the Brazilian food and beverage industry has been changing in recent years. Table 9.9 shows the concentration index CR_{10} from 1994 to 2000, indicating that the most important change was in the increased participation of MNEs, which in 1994 were only five and by 1999 had grown to eight. In this group, the only two firms with Brazilian capital are leaders in the meat market. The degree of concentration in processing, however, is relatively stable in contrast to other important segments of the food chain, such as distribution. Between 1994 and 2000 the ten largest supermarket chains doubled their share in the total turnover of the sector (Table A9.2). Increased concentration was an effect of acquisitions, espe-

TABLE 9.9. Concentration in the Brazilian food industry—CR$_{10}$

Firm	1994	Firm	1996	Firm	1999	Firm	2000
Nestlé*	5.42	Nestlé*	5.25	Nestlé*	6.01	Nestlé*	5.45
Copersucar	5.00	Copersucar	3.22	Ceval*	4.03	Bunge*	5.20
Ceval	3.53	Ceval	2.70	Sadia	4.03	Sadia	3.69
Santista*	3.28	Santista*	2.45	Cargill*	3.91	Cargill*	3.52
Sadia	2.89	Sadia	2.38	Perdigão	2.49	Perdigão	2.20
Frigobrás	1.68	Cargill*	1.92	Parmalat*	1.98	RMB*	1.68
RMB*	1.68	Perdigão	1.49	Santista*	1.98	Parmalat*	1.55
Perdigão	1.62	Parmalat*	1.47	Kraft Lacta*	1.33	Kraft Lacta*	1.09
Yolat*	1.51	Sadia Frigobrás	1.43	Arisco*	1.31	Fleishmann*	1.08
Cargill*	1.50	RMB*	1.29	Nabisco*	1.21	Aurora	0.87
Total (CR$_{10}$)	28.08	Total	23.60	Total	28.28	Total	26.32

Source: Exame Maiores e Melhores, 1995, 1997, 2000, 2001, in Farina, 2001, p. 13.
*MNEs

cially those by foreign capital, which doubled the presence of multinationals in this segment (Farina, 2002).

The vigor of competition in the food industry is reflected in the behavior of real prices in retailing, as shown in Figure A9.1. In the postmonetary stabilization period, an important drop in real prices was seen in the largest consumer center of the country, Greater Sâo Paulo.

The most important price reductions took place in processed foods, exactly where the multinational investments had occurred with the greatest intensity and where supermarkets are most important in distribution. Since demand was rapidly increasing in the same period, that behavior can be attributed to competitive pressure in the industrial segment and to the power of distribution in the determination of prices in urban areas.

As shown in Figure A9.2, among processed foodstuffs, meat and dairy stand out as leaders in real price reductions, with drops of 35 to 40 percent in the past six years. In dairy, the presence of foreign capital is growing, with an aggressive expansion policy from Parmalat (by means of M&As) and, among others, Nestlé and Danone. In meats, too, there was a rise of international capital, with acquisitions from Bunge and Doux, among others (Farina and Nunes, 2002).

Changes in Industry Allocation and Employment

In a period marked by several institutional and economic changes, it is difficult to define specific causal relationships. It should be noted that the entry of multinationals was concomitant with monetary stabilization and food market liberalization. As a whole, however, it is possible to argue that FDI did not represent only a transfer of ownership, since important changes in employment have been under way.

Data from RAIS (Annual Report of Social Information) permit an analysis of the behavior of formal employment in Brazil. We can verify an increase in the relative share of the food and beverage industry in industrial employment, notably in the most recent period. From 1988 to 1997, the share of the food industry grew considerably to reach 23 percent of the formal employment in Brazilian industries (see Table 9.10).

Nevertheless, the food industry was not immune to the restructuring undergone by all the industrial sectors and by the streamlining of

TABLE 9.10. Relative share in industrial Brazilian employment, by industrial sector.

Industry	1988	1993	1997	1988-1997 (% change)
Mineral extraction	2.3	2.2	2.0	−13.2
Nonmetallic min.	5.9	5.2	5.2	−11.3
Metal	10.3	9.2	9.7	−5.0
Mechanical	6.7	5.7	5.2	−22.7
Elect./commun.	5.1	4.7	3.7	−28.1
Transport mat.	5.6	5.7	5.5	−1.3
Furniture and wood	7.2	7.2	7.6	5.5
Paper and graphic	5.1	5.2	5.9	15.3
Rubber, tobacco, leather	7.1	6.7	4.7	−33.7
Chemicals	8.4	9.1	9.5	13.0
Textiles	14.6	14.8	13.4	−8.3
Shoes	5.0	6.0	4.4	−11.5
Food and beverage	16.6	18.4	23.0	38.8

Source: Relação Anual de Informações Sociais (RAIS), 1988, 1993, 1997.

costs, derived from the competitive pressures to which they had been submitted. Formal employment decreased in absolute terms. At the same time, as production grew, as shown on Figure A9.5, labor productivity gains increased. As of 1997, we see a stabilization in employment in the food industry, which suggests that the adjustment must have run its course (Figure A9.3).

Such productivity gains are consistent with the evolution of both average earnings and industrial production, shown in Figures A9.4 and A9.5. Average real earnings grew without interruption between 1992 and 1998 and remained relatively stable from then on. The index of industrial production, however, does not suffer the same modulation, which suggests that the increases in productivity continue, though they are no longer reflected in real earnings (Farina and Nunes, 2002).

What we could infer is that the growth of FDI is concomitant with growth in the production of processed foods and structural adjustment. This situation caused important changes in the pattern of competition, which induced falls in real prices, while at the same time it accelerated the release of new foodstuffs and intensified market segmentation. Therefore, it cannot be said that the entry of MNEs by means of M&As caused a mere transfer of ownership among Brazilian food firms.

CONCLUSIONS

The flow of FDI increased at the world level in the 1990s (see Chapter 4). Even though the greatest part of such investments continues to occur in industrialized countries, the flow destined for developing countries has increased still more in proportional terms. In this movement, Brazil has stood out in attracting investments to Latin America, being the largest recipient in the region. The Brazilian food industry has also attracted MNEs. The increase of FDI inflows to Brazil coincides with greater economic liberalization and decreased interference of the state in the economy. In addition to functioning as sources of new capital and as leaders in higher value-added activities, MNEs have promoted increases in international trade by facilitating the country's access to external markets and have contributed to greater dynamism in the domestic market.

Globalization, therefore, by increasing competition, forces firms to pursue greater efficiency to guarantee their markets at the world level. To understand such trends, one needs to explore the comparative and competitive advantages of each country in obtaining efficiency gains. Given the extent of recent Brazilian economic liberalization, acting in the national market does not guarantee the MNE, per se, significant market-share gains. Thus the releasing of new foodstuffs, for example, is vital for the growth and maintenance of the firm in the Brazilian market, because such a strategy confers greater competitiveness on companies, both national and multinational.

The Real Plan, by stabilizing the Brazilian economy, altered the institutional environment, favoring FDI. MNEs could better assess their possibilities for safe investments, attracted as they were by the vast potential of the domestic market. Furthermore, economic stability made it possible for a large portion of the population to enter the consumer market of several foodstuffs, from lower to higher value-added foods. Compared to saturated food markets in the home countries of MNEs, the growth rate of Brazilian demand has been a major factor of attraction. On the other hand, stabilization has increased the consumer's willingness to try new products, which has enhanced competition even more.

As main factors in attracting foreign capital to Brazil, we can pinpoint the following elements: (1) size of the Brazilian market; (2) interest in making Brazil an export base for MERCOSUR trade partners; (3) economic stability, in the initial period of the Real Plan; (4) fiscal incentives; (5) access to raw materials; and (6) low cost of labor.

Following Dunning's model, the enterprises join their personal advantages (O), such as the ownership of technology or their specific endowments (personnel, capital, organization), to the advantages of Brazil's location (L), such as quality and prices of inputs, specific resources, geographic location (the vast market in MERCOSUR), access to market knowledge, and mastery of the structure of distribution and supply of the local enterprises. Given such joint aspects (O + L), the firm could find it advantageous to internalize markets (I advantages) by means of direct production in the host country. As a consequence, uncertainty and transaction costs in the acquisition of inputs are reduced; the enterprises may avoid or better explore possible government measures, not to mention establishing price policies.

In analyzing data on employment, even with the reservation that the different industrial sectors use diverse technologies with different capital/labor ratios, we observe a considerable growth (39 percent in relative terms) of total employment in the food and beverage industry from 1988 to 1997, which is indicative of the relative increase in the share of this industry in the Brazilian manufacturing sector. This view is reinforced when we compare this evolution with the other industrial sectors that displayed negative or insignificant variations, for the most part. Thus, even with the predominant entry of MNEs by means of M&As, we note that increased FDI did not determine a mere transfer of ownership. M&As are the quickest way for an enterprise to enter a foreign country, increase its market share, and eliminate possible competitors; the strategy also provides it with efficiency gains by means of increases in scale and specialization of production. At the country level, increases in food production, concomitant with the reduction in the number of employees, encouraged labor productivity gains that were reflected in the increase in the average wage of the industry. These productivity gains are the result of major cost stream-lining, derived in turn from organizational changes and investments of physical capital, especially in automation (Farina and Nunes, 2002). Although such unequivocal efficiency gains cannot be attributed exclusively to the entry of MNEs, it was accelerated and magnified by them, because the foreign firms brought organizational and productive innovations to the competitive environment of the host country, in addition to inducing changes in the patterns of competition in the domestic market.

New product releases and new competitive strategies granted the Brazilian food industry greater dynamism, showing that FDI can provide a significant contribution to the technological and organizational development of the host country. As, in this case, multinationals are not exploring gains in countries with protected markets, competition intensifies in a global scenario, which increases the concern with efficiency gains and, consequently, with technological and organizational innovations. Moreover, multinationalization and consolidation in the retail segment reinforces the pressure for firms in upstream industries to become more efficient. Hence, the location of FDI is oriented as much by the competitive advantages of the host region as by the potential for innovation and productivity gain that it offers (Farina, 2002).

The MNE has considerable advantages over a national firm in releasing new products, because it has a real stock of products already released at the world level. Entering by means of M&A allows a rapid adaptation to local habits, facilitating the adaptation of these already existing products. Thus the national firm finds greater difficulties in keeping up with this pace of change because it requires more time to invest in new products that the MNE has often already released in other geographic markets.

Therefore, the changes in the institutional and politicoeconomic environment have made the Brazilian consumer food market highly attractive to investments by MNEs. Entering by means of M&A has been the best strategy. Such changes have altered the competitive environment, as suggested by the increase in industry concentration, with the growing participation of MNEs, and competition. This has in turn favored price stability and has encouraged firms to increase new product releases to avoid losses in their market share.

The difficulty in the analysis arises precisely because these modifications have occured in unison, making it hard to establish causal relationships. Hence, an important precaution must be taken. We cannot attribute these transformations exclusively to MNEs. That is, we have no way to verify if the national industries would have taken the paths discussed here even without the presence of multinationals, given that the domestic market potential in itself favors in large part alterations in the competitive environment (new product releases, price stability, and increased concentration). The presence of multinationals, however, is remarkable. This, without a doubt, at the very least accelerates these transformations, given the very characteristics of these enterprises (new technologies, market knowledge, international performance, vast dynamism, and competitive strategies). Entry alone does not guarantee multinationals a comfortable situation. In addition to the competition of several new multinational entrants, the greater degree of trade liberalization of Brazil and even national firms confront the MNE in the domestic market. Thus the dispute over supermarket shelves has gone on, causing firms to pursue more and more efficient strategies, at least to maintain their market share. This has increased the dynamism of the sector, even after the boom of multinational entry and the initial euphoria of the Real Plan passed.

APPENDIX

In this part of the chapter some data that support the previous analysis are presented. These include a list of the M&As in the Brazilian food and beverage industry from 1988 to 2000. A table with the market shares (in selected years) of the top ten supermarket chains in Brazil is also provided. A close look at these data gives a sense of the dynamics of the MNEs' entrance into the food industry in Brazil in the recent past.

TABLE A9.1. Main M&As in the Brazilian food and beverage industry, from 1988 to August 2001.

Date of operation	Acquired firm	Purchasing firm	Nationality of purchaser	% acquired	Value of transaction (US$ million)
12/15/00	Cafe Tres Coracoes SA	Elite Industries Ltd	Israel	100.00	41.0
12/12/00	Coop Central de Lacticinio do	Danone Group	France	100.00	118.895
11/15/00	Ipaussu SA Acucar e Alcool	Union das	France	47.50	39.0
10/24/00	Swift Armour Bordon-Meat	Bertin as	Brazil	100.00	36.652
08/31/00	Harald Industria e Comercio	Kerry Group PLC (Kerry Coop)	Ireland-Rep	100.00	20.0
05/30/00	Café União	Sara Lee Corp.	United States	100.00	215.828
05/24/00	Prenda SA	Chapeco Cia Industrial de	Brazil	100.00	15.0
03/31/00	Cia Antarctica Paulista	Cia Cervejaria Brahma	Brazil	100.00	368.594
03/01/00	Nitvitgov Refrigerantes SA	Embotelladora Andina SA	Chile	100.00	84.44
02/08/00	Arisco Produtos Alimenticios	Bestfoods	United States	100.00	752.0
02/04/00	Frigorifico Batavia SA	Perdigao Comercio e Industria	Brazil	51.00	11.871
12/28/99	Granja Rezende SA	Sadia as	Brazil	89.90	84.692
1999	Café Seleto	Sara Lee Corp.	United States	100.00	
1999	Café do Ponto	Sara Lee Corp.	United States	100.00	
11/26/99	Kraft Lacta Suchard SA	Adams do Brasil (Adams Spa)	United States	100.00	46.719
11/08/99	Chapeco Cia Industrial de	Grupo Macri	Argentina	100.00	213.0
03/03/99	Adria Produtos Alimenticios	Canale do Brasil SA (Canale SA)	Argentina	100.00	14.999
1999	Mococo	Royal Numico		N/A	N/A
10/23/98	Juiz de Fora Bottling Plants	Embotelladora Andina SA	Chile	100.00	119.993

311

TABLE A9.1 *(continued)*

Date of operation	Acquired firm	Purchasing firm	Nationality of purchaser	% acquired	Value of transaction (US$ million)
09/17/98	Refrigerantes do Oeste SA	Panamerican Beverages Inc	Mexico	100.00	36.358
09/03/98	Pullman Alimentos (Plus Vita)	Santista Alimentos (Bunge)	Argentina	100.00	84.999
07/21/98	Leitesol Industria e Comercio	Mastellone Hermanos SA	Argentina	93.80	14.126
04/29/98	Joanes Industria as	Archer Daniels Midland Intl	United States	100.00	64.003
03/03/98	Star & Arty Ingredientes	Kerry Group PLC (Kerry Coop)	Ireland-Rep	100.00	108.124
02/27/98	Industria Alimenticia Batavia	Parmalat do Brasil	Italy	51.00	200.0
02/06/98	Acquario Participacoes	Cirio-Polenghi-De Rica	Italy	100.00	24.976
02/06/98	Industrias Alimenticias Carlos	Cirio SpA	Italy	72.00	22.045
1998	VanMill	Plus Vita (Santista)			
1998	Frangosul	Doux S.A.		N/A	N/A
12/17/97	Sadia Concordia-Facilities (4)	ADM Exportadora e Importadora	United States	100.00	165.0
11/30/97	Baesa-Brazilian Operations	Cia Cervejaria Brahma	Brazil	100.00	155.001
10/31/97	Kibon SA Industrias Alimentici	Industrias Unilever	United Kingdom	100.00	930.0
10/23/97	Pepsi-Cola Engarrafadora,PCE	Cia Cervejaria Brahma	Brazil	100.00	150.0
07/25/97	Sadia Oeste-Meat Packing Plant	Friboi Alimentos Ltda	Brazil	100.00	13.863
04/28/97	Basilar as	Canale do Brasil SA (Canale SA)	Argentina	100.00	26.4
1997	Só Fruta	IANSA		N/A	N/A
1997	Ceval	Santista		N/A	N/A
1997	Matosul	Cargill		N/A	N/A

Date	Target	Acquirer	Country		
1997	Ceval	Bunge & Born		N/A	N/A
1997	Zabet	Canale		N/A	N/A
1997	Isabela	Canale		N/A	N/A
1997	Glencore (BR)	ADM			
07/02/96	Frescarini (LPC/Danone Group)	Pillsbury Co (Grand Met PLC)	United States	100.00	49.999
05/09/96	Cia Antarctica Paulista	Anheuser-Busch Companies Inc (joint venture)	United States	10.00	105.009
04/29/96	Industrias De Chocolate Lacta	Philip Morris Inc	United States	58.00	245.001
1996	Anderson Clayton	Unilever		N/A	N/A
1996	San Valentin	Cargill		N/A	N/A
1996	Naturalat	La Serenissima		N/A	N/A
1996	Visconti	Arisco		N/A	N/A
1996	Gumz Alimentos	Fleischman & Royal		N/A	N/A
1996	CIA. Produtos Pilar	Fleischman & Royal		N/A	N/A
1996	CCGL (CO-OP)	Avipal		N/A	N/A
1996	Bethania	Parmalat		N/A	N/A
1996	Aymoré	Danone		N/A	N/A
12/05/95	Pullman Alimentos	Plus Vita (Santista Alimentos)	Brazil	100.00	87.984
07/31/95	Usina são Jose (Votorantim)	Petribu Group	Brazil	100.00	53.475
06/09/95	Bestfoods	Unilever	United Kingdom	20.00	69.242
1995	Laticínios Avaré	Fleischan & Royal		N/A	N/A
1995	Moinho Campo Grande	Bung & Borg		N/A	N/A
1995	Pão Pullman	Bung & Borg		N/A	N/A

TABLE A9.1 (continued)

Date of operation	Acquired firm	Purchasing firm	Nationality of purchaser	% acquired	Value of transaction (US$ million)
1994	Agroeliane	Ceval		N/A	N/A
11/18/94	Adria Products Alimenticos	Quaker Oats Co	United States	100.00	100.0
06/14/94	Rio de Janeiro Refrescos SA	Sterling Pacific Co	Uruguay	100.00	120.0
05/10/94	Campineira de Alimentos	BSN as- Danone	France	49.00	34.875
05/28/93	Frutesp-Frozen Concentrated	Investor Group	Brazil	100.00	160.0
1993	Perdigão	8 Fundos de Pensão		N/A	N/A
1993	Cica S/A	Unilever		N/A	N/A
06/27/88	Beatrice Brazil (TLC Group)	Nestle SA	Switzerland	100.00	12.99
2002	Cervejaria Kaiser SA	Molson	Canada		30.0
N/A	Chapeco Cia Industrial de	Grupo Macri	Argentina		320.805
N/A	Perma	Embotelladora Andina SA	Chile		108.0
1999	Frangosul SA Agro Avicola	Doux	France		11.855
N/A	Plus Vita (Santista Alimentos)	Grupo Bimbo SA de CV	Mexico		63.5

Source: Thomson, data provided by BBV bank.
N/A = not available

TABLE A9.2. Concentration in supermarket chains.

Ranking		Supermarkets	1994	1996	1999	2000
2000	1999					
1	2	Pão de Açúcar[a,c]	6.5	7.4	12.9	14.1
2	1	Carrefour (French)[c]	9.4	10.4	13.1	14.1
3	4	Bom Preço (Royal Ahold—2000)[c]	2.4	2.6	4.4	4.5
4	3	SONAE (Portuguese)[c]	–	–	4.7	4.4
5	5	SENDAS	2.6	3.4	4.0	3.7
6	6	Wal-Mart (United States)[c]	–	–	1.6	1.8
7	7	Jerônimo Martins/SÉ (Portuguese)[b,c]	0.8	1.0	1.2	1.4
8	8	CIA Zafari	0.9	1.3	1.1	1.1
9	9	G. Barbosa & CIA LTDA	0.5	0.6	0.8	0.9
10	10	Cooperativa de Consumo	–	–	0.8	0.8
		Top ten market share (%)[d]	24.3	28.4	44.6	46.8

Source: Associação Brasileira de Supermercados (ABRAS) <www.abrasnet.com.br>.
[a]40 percent is owned by the French Cassino.
[b]In negotiation with the Dutch Royal Ahold.
[c]MNE affiliates.
[d]Reflects only the sum of the supermarkets in Brazil that were ranked in the top ten.

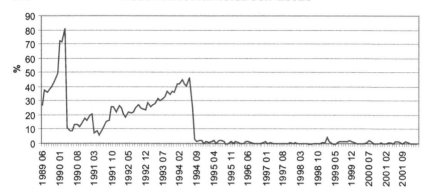

FIGURE A9.1. General price index from June 1989 to March 2001 (IGP-D!, FGV). (*Source:* Getúlio Vargas Foundation, <www.ipeadata.gov.br>.)

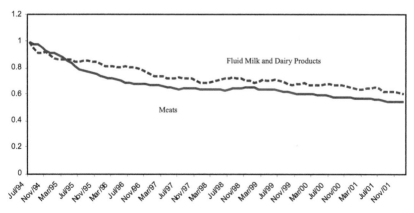

FIGURE A9.2. Index of consumer real prices (basis: July 1994 = 1.00): Fluid milk and dairy products, meats. (*Source:* FIPE—Fundação Instituto de Pesquisas Econômicas [Economic Research Institute] <www.fipe.com>.)

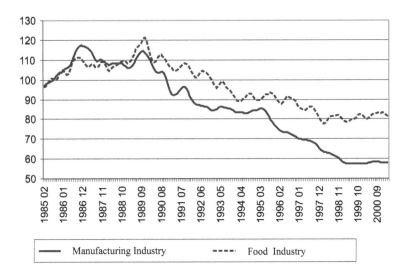

FIGURE A9.3. Employment in the manufacturing industry and in the food industry (three-year moving average, basis 1.00 = mean 1985). (*Source:* Farina and Nunes, 2002, based on basic data from IBGE.)

FIGURE A9.4. Average real earnings in the manufacturing industry and in the food industry (three-year moving average, basis 1.00 = mean 1985). (*Source:* Farina and Nunes, 2002, based on data from IBGE.)

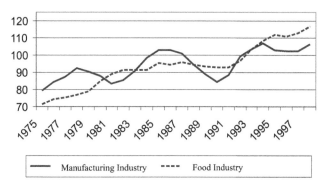

FIGURE A9.5. Production of the manufacturing industry compared with the food industry (moving average, basis 1991 = 100). (*Source:* Farina and Nunes, 2002, based on data from IBGE <www.ibge.gov.br>.)

NOTES

1. The Real Plan was a monetary stabilization plan implemented in July 1994. It brought down the inflation from 2708 percent per year (IGP-DI, Getúlio Vargas Foundation, in 1993) to less than 10 percent per year. The accumulated real inflation since initiation of the plan to April 2002 reached 124 percent. The plan played an important role in increasing the per capita real income.

2. This value refers to the net direct investment, with credit equivalent to US$40.290 billion and debit equivalent to US$7.511 billion. Brazil's Central Bank uses the International Monetary Fund's (IMF) payments balance methodology. See <www.bacen.gov.br>.

3. The extended MERCOSUR is composed of Brazil, Argentina, Uruguay, Paraguay, Chile, and Bolivia.

4. This aspect is better explored by Caves (1996).

5. Brazilian inflation reached 80 percent per month prior to 1994. The Real Plan was the fifth stabilization plan since the mid-1980s.

6. Research presented by the *Gazeta Mercantil Latino-Americana,* August 7-13, 2000, p. 8.

7. During the period of hyperinflation, people received their salaries and went straight to the supermarket so they would not lose purchasing power. The prices were readjusted every day.

BIBLIOGRAPHY

Arroyo, Gonzalo, Rama, Ruth, and Rello, Fernando (1985). Agricultura y Alimentos en América Latina—el poder de las transnacionales. Universidad Nacional Autónoma de México, Programa Universitario de Alimentos. México, D.F., Instituto de Cooperación Iberoamericana (ICI) Madrid.

Baumann, Renato (Ed.) (1996). *O Brasil e a Economia Global.* Rio de Janeiro, Brazil: Editora Campus.

Belik, W. (1994). Food Industry in Brazil: Toward a Restructuring? Research Paper number 35. Institute of Latin America Studies, University of London.

Belik, W. (1995). Agroindústria e Reestruturação industrial do Brasil: Elementos para uma Avaliação. In P. Ramos and B. Reydon (Eds.), *Agropecuária e Agroindústria no Brasil: Ajuste, situação atual e perspectivas* (pp. 107-123). Campinas, Brazil: Edição Abra.

Bolling, Christine, Neff, Steve, and Handy, Charles (1998). U.S. Foreign Direct Investment in the Western Hemisphere. Washington, DC: Economic Research Service, USDA.

Bonelli, Regis (2000). Fusões e Aquisições no Mercosul. IPEA-RJ, Texto para discussão no. 718, April.

Brazilian Institute of Geography and Statistics (IBGE). (2002). Dados macro-econômicos. Online: <www.ipeadata.gov.br>, accessed May 4.

Canuto, Otaviano (1994). Abertura Comercial, estrutura produtiva e crescimento econômico na América Latina. *Revista do Instituto de Economia da Unicamp, Economia e Sociedade,* (December): 43-64.

Carmo, Maristela Simões do (1996). *(Re)estruturação do Sistema Agroalimentar no Brasil: A diversificação da demanda e a flexibilidade da oferta.* Coleção Estudos Agrícolas 5. São Paulo, Brazil: Instituto de Economia Agrícola.

Caves, Richard (1996). *Multinational enterprise and economic analysis,* Second edition. Cambridge Surveys of Economic Literature. New York: Cambridge University Press.

Cebrap (1997). Concentração e Centralização de Capitais na Indústria de Alimentos Brasileira. Anexo ao relatório final de pesquisa do projeto: Democracia e poder econômico: A legislação antitruste brasileira diante dos processos de concen-tração e centralização de capitais à escala mundial. Relatório de Pesquisa, Centro Brasileiro de Pesquisa.

Chesnais, François (1996). *A mundialização do capital,* First edition (Portuguese). São Paulo, Brazil: Editora Xamã.

Connor, John M. and Schiek, William A. (1997). *Food processing: An industrial powerhouse in transition,* Second edition. New York: John Wiley and Sons.

Coutinho, Luciano and Ferraz, João (Eds.). (1994). *Estudo da Competitividade da Indústria Brasileira.* Campinas, Brazil: Editora da Universidade de Campinas—Unicamp.

Dunning, John H. (1988). The new style multinationals, circa the late 1980s and early 1990s. In J. H. Dunning (Ed.), *Explaining multinational production.* London: Unwin Hyman.

Dunning, John H. (1995a). *The globalization of business.* New York: Routledge.

Dunning, John H. (1995b). *Multinational enterprises and the global economy.* Boston: Addison Wesley.

Economic Research Service (ERS/USDA) (2001). Database. Available at <www. ers.usda.gov>.

Farina, E.M.M.Q. (1999). Challenges for Brazil's food industry in the context of globalization and Mercosur consolidation. *International Food and Agribusiness Management Review* 2(3/4): 3-4.

Farina, E.M.M.Q. (2001). Downstream structural changes in the Brazilian food and agribusiness system: The case of dairy business. Unpublished paper.

Farina, E.M.M.Q. (2002). Consolidation, multinationalization, and competition in Brazil: Impacts on horticulture and dairy products systems. *Development Policy Review* 20(4): 441-457.

Farina, E.M.M.Q., Azevedo, Paulo Furquim de, and Saes, Maria Sylvia M. (1997). *Competitividade: Mercado, Estado e Organizações*. São Paulo: Singular.

Farina, E.M.M.Q., Giordano, S., Viegas, C., and Farina, T.M.Q. (2001). Agri-food business in Brazil. In C. Brown and P. Rennart (Eds.), *Doing business in Brazil* (pp. 49-52). Sydney: Department of Foreign Affairs and Trade and Austrade (Australian Trade Commission), Commonwealth of Australia.

Farina, E.M.M.Q. and Nunes, R. (2002). A evolução da cadeia agro-alimentar e a redução de preços para o consumidor: os efeitos da atuação dos grandes compradors. Relatório de pesquisa, Convênio CEPAL/PENSA/USP, April.

Farina, E.M.M.Q. and Saes, Maria Sylvia M. (1996). Food industry in Mercosur: Many challenges and big opportunities. Paper presented at the international seminar Food, Agriculture and Agribusiness: Future Challenges and Opportunities, Royal Agriculture College, UK, September 27.

Faveret, Paulo (1999). *Fusões e aquisições no setor de alimentos*. Informe Setorial: Agroindústria. Brazil: BNDES, FINAME, BNDESPAR.

Gazeta Mercantil—Latino Americana. Various issues.

Glass, Amy Jocelyn and Saggi, Kamal (1999). Foreign direct investment and the nature of R&D. *Canadian Journal of Economics* (Revenue Canadienne d'economie) 32(1): 92-117.

Humphrey, John and Oetero, Antje (2000). Strategies for diversification and adding value to food exports: A value chain perspective. Institute of Development Studies, University of Sussex. United Nations Conference on Trade and Development (UNCTAD), November 14.

Laplane, M. F. and Sarti, F. (1997). Investimento Direto Estrangeiro e a Retomada do Crescimento Sustentado nos anos 90. *Rev. Economia e Sociedade* 8(June): 143-181.

Martinelli Orlando, Jr. (1997). As tendências da Indústria de Alimentos: Um Estudo a partir das Grandes Empresas. Doctoral thesis. Instituto de Economia da Universidade Estadual de Campinas.

Negri, João Alberto de (1997). Impactos das Multinacionais na Reestruturação da Indústria: Uma proposta metodológica. Instituto de Pesquisa Econômica Aplicada (IPEA), Texto para discussão nº 474, May.

Rama, Ruth (1992). Investing in food. Paris: OECD (Development Centre of the Organisation for Economic Co-operation and Development).

Rama, Ruth (1998a). Growth in food and drink multinationals, 1977-94: An empirical investigation. *Journal of International Food and Agribusiness Marketing* 10(1): 31-52.

Rama, Ruth (1998b). Tasa de beneficio e innovación en los grupos estratégicos de la indústria alimentária internacional. *Cuadernos de Economia y Dirección de la Empresa* 4: 285-300.

Rama, Ruth (1999). Industria agroalimentaria: Innovación y globalización. Comercio Exterior—Banco Nacional de Comercio Exterior. *Desarrollo territorial y Globalización*, México, 49(8).

Relação Anual de Informações Sociais (RAIS) (1988). CD-ROM. Brazil: Ministério do Trabalho e Emprego (MTE).

Relação Anual de Informações Sociais (RAIS) (1993). CD-ROM. Brazil: Ministério do Trabalho e Emprego (MTE).

Relação Anual de Informações Sociais (RAIS) (1997). CD-ROM. Brazil: Ministério do Trabalho e Emprego (MTE).

United Nations Conference on Trade and Development (UNCTAD) (1999). *Foreign direct investment and the challenge of development.* World Investment Report. New York and Geneva: United Nations.

United Nations Conference on Trade and Development (UNCTAD) (2000). *Cross-border mergers and acquisitions and development.* World Investment Report. New York and Geneva: United Nations.

United Nations Conference on Trade and Development (UNCTAD) (2001). *Promoting linkages.* World Investment Report. New York and Geneva: United Nations.

United Nations Conference on Trade and Development (UNCTAD) (2002). *Transnational corporations and export competitiveness.* World Investment Report. New York and Geneva: United Nations.

Vegro, Celso Luís Rodrigues and Sato, Geni Satiko (1995). Fusões e Aquisições no Setor de Produtos Alimentares. *Informações Econômicas* SP 25(5): 9-21.

Viegas, Cláudia Assunção dos Santos (2002). Empresas Multinacionais na Indústria Brasileira de Alimentos. Master's thesis. Economics Department, University of São Paulo.

Index

Page numbers followed by the letter "t" indicate tables; those followed by the letter "f" indicate figures; those followed by the letter "b" indicate boxed text; and those followed by the letter "n" indicate notes.

Acquisitions, 5t, 29t-31t. *See also*
    Mergers and acquisitions (M&A)
    in Brazil, 283, 285
    and FDIs, 150
ADM (Archer Daniels Midland), 19
Advertising expenditure, and
    internalization of exchanges,
    195
Agrodata, 2, 13, 17, 28, 80
Asian conglomerates, 258-259, 270-271
Australia and New Zealand
    food production firms, 221t-223t
    food sector history, 226-231

Biotechnology patents, 122
    Japan, 135
Brazil
    agrofood sector, 283-284
    bilateral FDI, 290t
    consumer market, 298, 299t, 300t,
        301-303
    employment, 305, 306t, 317t
    food and beverage industry changes,
        303-305, 304t, 316t
    mergers and acquisitions, 283, 285,
        295-298, 295t, 296t, 311t-315t
    productivity gains, 306t, 318t
Bulk commodities trade, 149
By-product processing, 132

Cadbury Schweppes, 19
Canada, as host country, 289

Carrefour (France), retail sector, 273,
    278n.2
Chaebols, South Korean multisectorals,
    257
Charoen Pokphand (CP) Group,
    Thailand, 256, 257, 260-265,
    273, 275, 276, 277
Chemistry, food-related innovation,
    133t, 134-135
China, agrofood market, 276
Coca-Cola, 15, 132
Coca-Cola Amatil (CCA), 220,
    236-242
Complementarily, outward FDI and
    exports, 166-167
ConAgra, 16, 19, 132
Consolidation, U.S. food MNEs, 22-23,
    23f
Core business activities, 15, 16, 19
    FBMs, 75
    top 100 food MNEs, 59t-63t
Country spread, *xvii*
Country-specific approach, third world
    countries, 256
Country-specific resources, 122
"Cultural distance," and FDI, 161
Curvilinear model, diversification
    success, 78

Dairy products, emerging markets, 16
Deregulation, impact in Australia and
    New Zealand, 219, 220

Design patents
    F&B multinationals, 118f, 136
    foreign locations, 123
Divestitures, Unilever, 12b
Downstream industries, and
        diversification, 78
Dunning, John H., 166, 168, 170, 220,
        224, 225, 255, 256, 289, 292,
        293, 295b, 308

Eastern and Central European (ECE)
        market, 26
Eclectic framework/paradigm,
        international production, 220,
        292
Economic development, and FDI, 162
Economies of scale, in intrafirm trade,
        194
Emerging markets, dairy products, 16
European Union (EU) countries,
        agricultural politics in, 27. *See
        also* Western Europe
Exchange rate
    and FDI, 160
    and food exports, 166
Export-enhancing FDI, 256
Export-oriented industrialization (EOI),
        258
Exports
    and FDI, *xix,* 166
    and intrafirm trade, 192

Factor movements, in FDI, 156
Financial incentive, and FDI, 168
Firm
    internationalization paths, 220
    role of in foreign trade, 191
Firm-specific internationalization,
        antipodean food firms, 220,
        224-226
Fonterra, 220, 242-248

Food and beverage (F&B) industry
    growth rate, 73
    processing, 115
    stable sales, 86, 87t
Food and beverage multinationals
        (FBM)
    diversification patterns, 82-83,
        84t-85t, 87t, 88t, 91t, 93t
    expansion, 74
    innovation, 115-116, 121-122
    nonfood technology, 132, 133t
    patents, 117-119, 132, 133t
    top 100, 80, 107
Food and drink
    agroprocessing, *xvi*
    marketing innovation, *xvii-xviii,* 121
Food-processing capital, 22
Foreign affiliates
    products of, 152, 153f
    sales by, 153, 154f
    sales of, 151, 152f
Foreign direct investment (FDI)
    Brazil, 284, 285, 308-309
    definition, 149
    determinants of, 158-161
    in developing countries, *xxi*
    future research, 162-163
    growth, 165
    inbound, 154, 158
    Latin America, 286
    location, 168-169
    modes of entry, *xvi*
    motivation for, 167-168
    multinational enterprises, 156, 158
    outbound, 154, 158
    recent trends, 150-155
    in Southeast Asia, *xx*
    and trade, *xviii-xix,* 150, 166,
        169-171
    in TWMNEs, 254-257
    U.S. outflows, *xviii,* 151, 151f,
        154-155, 155f, 291t
    U.S.-based food processors, 174,
        175t
    western hemisphere, 291t

Foreign location patents, 120, 122-123, 124t
in FBMs, 74
and FDI, 150
Foster's Brewing Group (FBG), 231-236
France
food exports, *xix,* 199t
internal agrofood market, 200-201
intrafirm trade ("Globalisation")
survey, 192, 198
France, trade balance, 199, 199t, 200t
French agrofood IFT
export volume, 212-214, 213t
factors, 210t
modeling, 205, 207-208, 210-211
selling prices, 211, 212t
variables, 209t
French Ministries of Industry (SESSI),
Agriculture (SCEES), 198
Frito-Lay, 15

General Agreement on Tariffs and
Trade (GATT), 23-24
Geographical dispersion intensity
(GDI), 13, 19, 69n.8
top 100 food MNEs, 64t-68t
Geographical diversification, 79-80,
102, 103-104
controversy, 75-76
and firm growth, 94, 97t-98t, 101t,
102t
top 100 FBMs, 83, 84t-85t, 87, 88t
Geographical expansion, MNEs, 10-14
Geographical globalization indicator
(GGI), 13
top 100 food MNEs, 64t-68t
Geographical zones, Agrodata, 13,
69n.6
Germany, FBM diversification, 74
Global brand names, 10, 12
Global firm, expansion, *xvii*
Globalization, *xv*
agrofood systems, *xvi*
Brazilian food industry, 307

Globalization *(continued)*
developing country impact, *xxi*
food-processing, 12
and harmonization, 22
and intrafirm trade (IFT), 214
MNEs, 3-10, 7f
and technological development, 120
third world multinational enterprises
(TWMNEs), 253-254
Globalizing drivers, 3
Greenfield investments, 12-13, 69n.5
Gross domestic product (GDP), and
food exports, 166
Groupe Danone, 12b, 15, 19, 22
Growth diversification, F&B industry,
74
Growth rate, FBMs, 90, 93t

Harmonization, in globalization, 22
Headquarters country, top 25, 155, 157t
Heckman procedure, 206b-207b
Home zones, 18
Human capital, and internalization of
exchanges, 195

Indonesia, Salim Group, 265-269, 275
Industrial diversification, 76-79,
103-104
definition, 76
and firm growth, 94, 99t, 100t, 101t,
102t
measures of, 81
top 100 FBMs, 83, 84t-85t, 89
Industry-specific influences, antipodean
food firms, 220, 226-231
Industry-specific study, FDI, 162
Input prices, and FDI, 161
Intellectual property rights, and FDI,
159
Interest rate, and FDI, 161
Internalization advantage (I), in OLI
analysis, 17, 168, 292, 294

National Institutes of Statistics
(INSEE), France, 198
Nestlé, 8, 12b, 15, 18, 22, 132
New products, and internalization of
exchanges, 195
Newly industrialized countries (NICs),
258
North American Free Trade Agreement
(NAFTA), *xv,* 289

OLI (ownership, location, and
internalization) theory, 168,
292-294, 295b
Operations-extending FDI, 256
Organisation for Economic Co-operation
and Development (OECD), 121
Outbound FDI and export
data, 174-177, 176t, 178t
empirical model, 172-174
future research, 187
results, 177, 179t, 180, 181t, 182t,
182-183, 184t, 185t, 185-186
Ownership advantage (O), in OLI
analysis, 17, 168, 292,
293-294

Panel-data econometric model
firm growth, 94-96
results, 96
Patent innovation
competition, 138
FBMs, 117
growth, 118f, 119t
home-region/host-region
knowledge, 130
population influence, 131-132, 137
sample data, 142-143
Patent statistic analysis, 116, 143n.1
Penrosian theory, *xvii*
PepsiCo, 15, 16
Pernod Ricard, 19
Philip Morris, 18
Plant closures, 8, 9t

Positioning
top 100 food MNEs (1988), 20f
top 100 food MNEs (2000), 21f
Probiotics, 10
Processed food product trade, 149
Product diversity, 77
Protection policies, and FDI, 161

Ranking, top 100 food MNEs
1974, 54t-58t
1988, 48t-53t
2000, 42t-47t
Real Plan, Brazil, 283, 284, 307, 318n.1
Refocusing, 18, 59t-63t
Regionalization processes, *xv*
Related diversification, 75
Research and development (R&D)
facilities, *xv, xviii*
funding, 116
internationalization, 122-127
and intrafirm trade, 195
organization, 135
priorities, 132, 133t
U.S. companies, 127-129
Resource-based theory, diversification,
77
"Rest of the world, the," 24-25, 70n.20
Restructuring, 5t, 6t, 16, 38t-41t
Retail sector, Asian agrofood
conglomerates, 271-274,
278n.2
Retail trade, food industry investment
in, 151-152, 154f
Risk, and FDI, 162

Salim Group, Indonesia, 257, 265-269,
273, 275, 277
Second-tier firms, *xxi*
Sectoral globalization indicator (SGI),
18, 19, 70n.15
top 100 food MNEs, 59t-63t
Sell-offs, 11t, 35t-37t

Single market, European agrofood
    industry, 201-205
Single product MNEs, 10
Single-business companies, versus
    diversified, 76
Size, FBMs, 90, 91t-92t
Small and medium-sized enterprises
    (SMEs), 8
Southeast Asian agrofood
    multinationals, 269-271
    retail sector, 271-274
Strategic relatedness, 77

Taxation, and intrafirm trade, 197
Technoglobism, *xv*, 120
Technological content, and
    diversification, 79
Tesco (UK), retail sector, 273, 278n.2
Thailand, Charoen Pokphand Group,
    260-265, 275
Third world multinational enterprises
    (TWMNEs), 253-254
    recent studies, 254-257
    state influence, 274-275
Three-bond advantage system, 17
"Tiger" economies, 258
Tobacco
    foreign sales, 152, 153f
    industry diversification, 75
Trade barriers, and foreign production,
    158
Trade liberalization, Australia and New
    Zealand, 219
Trade volume, and FDI flows, 150f
Transportation, and foreign production,
    158
Triad (North America, Japan, Western
    Europe) market, 3, 16, 22

Unilever, 8, 12b, 18, 22, 26, 121, 132
United Kingdom, world food oligopoly,
    25f, 25-26
United Nations–Standard Industrial
    Classification (UN–SIC), 117
United States
    acquisitions and takeovers, 5t
    distribution changes, 22, 24f
    food-processing downsizing, 7
    foreign affiliate sales, 151, 152f
    mergers and acquisitions, 6t
    sell-offs and spin-offs, 11t
United States Department of
        Commerce, intrafirm trade,
        192
Unrelated diversification, 71
Upstream industries
    and diversification, 78
    and innovation, 134
Utility patents
    F&B multinationals, 118f, 136
    foreign locations, 123

Vertical relatedness, 78

Western Europe
    acquisitions and takeovers, 5t
    composition changes, 25-27
    distribution food MNEs, 25f
    food-processing downsizing, 7
    mergers and acquisitions, 6t, 10
    R&D globalization, 128-129
    sell-offs and spin-offs, 11t
    shrinking market factors, 8, 14
Wholesale trade, food industry
        investment in, 152
Workforce, and FDI, 168

Printed and bound by CPI Group (UK) Ltd, Croydon, CR0 4YY

23/10/2024

01777693-0001